Students' Attitudes, Perceptions, and Expectations Toward Instructional Technology in Higher Education: A Diffusion of Innovations

Students' Attitudes, Perceptions, and Expectations Toward Instructional Technology in Higher Education: A Diffusion of Innovations

By

Mamie Lewis Johnson, Ed.D.

iUniverse, Inc.
New York Bloomington

Students' Attitudes, Perceptions, and Expectations toward Instructional Technology in Higher Education
A Diffusion of Innovations

iUniverse books may be ordered through booksellers or by contacting:

iUniverse
1663 Liberty Drive
Bloomington, IN 47403
www.iuniverse.com
1-800-Authors (1-800-288-4677)

ISBN: 978-1-4401-7629-6 (sc)
ISBN: 978-1-4401-7630-2 (ebk)

Printed in the United States of America

iUniverse rev. date: 06/07/2011

Contents

Acknowledgements

I would like to thank all the wonderful people who gave me the courage and motivation to stay the course and not give up. First and foremost, I give thanks to God, my heavenly Father who gave me life, salvation, and wisdom to trust in Him. My sincere gratitude and appreciation to Dr. Edward Hill, Jr. for his dedicated advice and support of this book. I am indebted to Dr. Everett M. Rogers, the theorist behind diffusion of innovations, who gave me encouragement and was instrumental in this visionary project. I am truly inspired by his scholarly work on the diffusion of innovations, and I am glad to have met him. Offering sage advice behind the scenes was Dr. Gary Confessore for his wisdom and academic support. He encouraged me to "travel the road not taken" because he saw the potential in me that would one day come to fruition. For that kindness and soaring personality, I thank him. Thanks to a special friend and counselor, Dr. Orren Rayford, for his endearing friendship, guidance, and motivation. And last, to Rev. Dr. J. Allen Lewis (deceased), my beloved brother who supported me with a mental attitude of courage and devotion. I shall always be grateful to my entire family for their love and continued support.

About the Author

Dr. Mamie L. Johnson is an assistant professor of English in the Department of English and Foreign Languages at Norfolk State University. Previously she served as Dean of Academic Support Services, Head of the Department of English and Foreign Languages, and Associate Dean of Liberal Arts at Livingstone College, Salisbury, NC, where she conducted research on undergraduate students in the undeclared curriculum, retention of underrepresented students, and student engagement at four-year institutions. She earned her Ed.D. in higher education administration from The George Washington University, an M.A. degree in Communication with a minor in public relations and a B.S. degree in Administration Systems Management from Norfolk State University.

Johnson's interests include instructional technology, retention and assessment, student engagement, and academic excellence.

Introduction

The project on which this book is based follows the intent to describe and analyze the adoption patterns and characteristics of students' attitudes, perceptions, and expectations toward their new learning environment, associated with the introduction to instructional technology. Based on Everett M. Rogers's (1995) theory of the diffusion of innovation, the study classified students into five categories according to the behavioral characteristics they exhibited toward instructional technology. Based on specific characteristics, students were grouped as innovators, early adopters, early majority, late majority, or as laggards. For the adopter categories, Rogers's innovation-decision process was used to analyze the difference between early adopters and mainstream adopters, as well as to examine the issues surrounding the integration of instructional technology for teaching and learning in higher education. The first task was to examine the characteristics of "innovators," those who are venturesome, risk takers, and trend setters. The second task was to identify the characteristics of "early adopters" as those who are independent and self-reliant individuals in the adoption of instructional technology.

The relationship between early adoption and autonomous learning is explored in chapter two. In the third task, the investigator classified the "early majority" as those who are deliberate and provide interconnectedness in the system's networks. The fourth task was to classify the "late majority" as those who are skeptical or cautious individuals. The final step was to classify the "laggards" as those who are the traditional last to adopt an innovation. The five categories of adopters and how they perceive instructional technology as a pedagogical approach to learning serve as the purpose of this book.

Preface

Colleges and universities are charged with taking a proactive stance to meeting the educational needs of students by preparing them to meet the challenges of a global society in the 21st century. The purpose of this study was to describe and analyze the adoption patterns and behavioral characteristics of students' attitudes, perceptions, and expectations toward their new learning environment, associated with the introduction to instructional technology. This correlational study builds and extends on Everett Rogers's (1995) framework of the diffusion of innovations, and students' responses were classified into five adopter categories: innovators, early adopters, early majority, late majority, and laggards. The study was conducted at Old Dominion University, Norfolk, Virginia and The University of Virginia, Charlottesville, Virginia with 171 volunteer students taking instructional technology courses. Factors influencing the adoption of instructional technology were time, prior exposure or knowledge, and credit hours of experience. Greater knowledge and use of instructional technology was associated with more favorable attitudes and expectations toward instructional technology. The study suggests that students are highly motivated in their new learning environments and expect educators to incorporate more instructional technology into their pedagogical methods. The study further suggests that instructors dealing with change in educational settings should consider students' behavioral patterns of acceptance when attempting to understand students' adoption of innovativeness in a new learning environment.

CHAPTER 1

Integrating Instructional Technology into the Classroom

One of the most fundamental challenges facing higher education today is the diffusion of integrating the use of instructional technology into the classroom. Instructional technology gives new meaning and new dimensions to the traditional classroom and the way in which students learn. For purposes of this book, ***instructional technology*** (IT) is defined as the delivery of instruction with the aid of computer software and hardware as tools to enhance the teaching and learning process. Although it is a complex and integrated process involving people, procedures, ideas, devices, and organization for analyzing problems, instructional technology has been defined by professionals of the Association for Educational Communications and Technology (AECT) and accepted in the field as "the theory and practice of design, development, utilization, management, and evaluation of processes and resources for learning" (Seels and Richey, 1994). The authors assert that both the theory and practice components of the 1994 definition incorporate the use of procedural models. Another definition for instructional technology is "the systemic and systematic application of strategies and techniques derived from behavior and physical science concepts and other knowledge to the solution of instructional problems" (Gentry, 1995, p. 7). As a result, the term instructional technology is construed in a very broad sense as referring to any systematic approach to teaching and learning with the aid of machine-based technology as a component of audiovisual aids such as PowerPoint, Blackboard, teleconferencing, web base instruction, internet, or world wide web connections, just to name a few.

Other specialists who have been involved in the field of instructional technology have defined it in several ways. Ralph Lambert (1999) defines it as "any point where technology joins instruction and any place where faculty employ technology to aid them in the completion of their instructional tasks" (p. 24). He asserts that technology must add value to the content and enhance the learning process and not just entertain students. Lambert argues that instructional technology begins with the most basic software programs that a faculty member might employ to aid in class preparation, and it also includes software and/or hardware that students might use to enhance their critical thinking skills while learning the course materials.

Still, others like Dick and Carey (1996) have provided models for instructional strategies that might be applied to the proper selection of Internet-based delivery systems. The strategies they suggest retain the learning objective as the primary influence for media selection. The authors suggest using a computer-based delivery system as a favorable medium for teaching and developing intellectual skills. Instructional technology begins with the most basic software programs that instructors use to aid in their class preparation and presentations. It may include but is not limited to graphics, the Internet, web page development, web course presentations, e-mail messages or correspondence, word processing, CD-ROM's, videos, or distance learning modalities and related technologies.

However, the most intriguing definition that best relate to this study includes Knezevich and Eye's (1970) definition, which states it is "an effort with or without machines, available or utilized, to manipulate the environment of individuals in the hope of generating a change in behavior or other learning outcome (p. 16). Higher education's main audience for this technological change is students. As primary stakeholders, students are no longer viewed as dependents or those who succumb to the powers of the professors; rather, they are life-long learners who are challenged to be competitive consumers in a global society. Furthermore, students are a means to the adoption of innovative advancements to technology. As a result, educators will seek to use the diffusion of instructional technology as a marketing effort to increase the adoption of innovative practices for future generations. Educators, faced with the growing realization that higher education has suffered from a lack of utilization of technology, are beginning to turn to the diffusion of instructional technology in an effort to increase the adoption of innovative technologies in the academic realm.

In order to fully capitalize on instructional technology as a cognitive tool for enhancing teaching and classroom content, educators must embrace this technological innovation as part of the teaching and learning process. The innovation may have an effect on the cognitive development and behavioral characteristics of students as well as produce fundamental changes to the organizational structure of the classroom content. Since it will be increasingly important for all students to learn, it is imperative that higher education institutions continue to play a significant role in the production and transmission of knowledge to the students. Students in higher education will be challenged to function in their learning environments with the ability to adopt instructional technology as an innovative change of integrating computer technology into the learning process.

This study focused on a sample of 171 undergraduate students at two public four-year institutions in southeastern Virginia. The investigator sought to determine if there was a relationship among students' attitudes, perceptions, and expectations toward instructional technology associated with an introduction to instructional technology. This study investigated the diffusion of instructional technology from the students' perspectives and should contribute greatly to the research of instructional technology in higher education. Results from the study may provide valuable information for future researchers to determine measures for improving instructional technology methods as dictated by students' attitudes, perceptions, and expectations toward instructional technology. The study results may also reveal the impact of technology on learning in post-secondary education and may contribute extensively to the diffusion of innovative literature.

Existing research shows that instructional technology has been under constant criticism for the past three decades. As a matter of fact, this criticism has penetrated the doors of higher education, causing a sharp decrease in the number of studies on this subject. An explanation for this decline may stem from the fact that other literature expressed only opinions and discussions about the use of instructional technology in the classroom (Clark, 1983; Croft, 1994; Denk, Martin, & Sarangarm, 1993). Furthermore, according to Bjork (1991), many of the research methods used to evaluate educational technology were ineffective. In addition, research on the use of instructional technology only addressed the type of technology used rather than the impact it had on students' behavioral characteristics as learners. Some studies examined several types of software and their effectiveness but did not identify behavioral patterns expressed mainly by students' attitudes, perceptions, and expectations toward instructional technology associated with their new learning environments. The use of instructional technology in the classroom is changing students' behavioral characteristics toward achieving instructional objectives. Knowing which methodological strategies that involve instructional technology to be perceived as being effective or ineffective could be valuable information for instructors. In addition, it could be an added advantage for students and administrators in colleges and universities nationwide. The problem being investigated is to determine whether or not there is a correlation among students' attitudes, perceptions, and expectations toward instructional technology, associated with innovativeness as an independent variable in this study.

While computer technology may be radically enhancing instructional methods and promoting enormous changes in curriculum development, the traditional methods of classroom teaching are still very much a part of the average professor's delivery mode in higher education. Therefore, this study investigated whether or not students were adopting the diffusion of innovation in instructional technology methods in their new learning environment, or whether they were rejecting the innovation. The study analyzed and evaluated the effectiveness of instructional technology as students were placed in a new learning environment based on their attitudes, perceptions, and expectations toward their new learning environment when they were introduced to instructional technology. This investigation also examined changes in students' attitudes, perceptions, and expectations toward technology in two public four-year universities where typically instructional technology was used as an integral part of teaching and learning.

History of Diffusion

The early days of diffusion research date back to the humble beginnings of a French lawyer and judge named Gabriel Tarde, the forefather of sociology and social psychology. Rogers asserted that Tarde was ahead of his time in thinking about diffusion. As Rogers (1995) cited in his book, *Diffusion of Innovations*, "At the turn of the 20th century, Gabriel Tarde kept an analytical eye on trends in his society as represented by the legal cases that came before his court"(p. 40). He was a pioneer in making amazing discoveries of identifying the adoption or rejection of innovations by those whose behaviors he observed as a crucial outcome variable.

According to Rogers (1995), "Tarde observed that the rate of adoption of a new idea usually followed an S-shaped curve over time" (p. 40). The purpose of Tarde's observations of people was to learn why some innovations would spread abroad while others would not last for more than a few days. Tarde observed certain generalizations about the diffusion of innovations in society that he called "the laws of imitation" (p. 40). What Tarde once called "imitation," Rogers now refers to as the adoption of an innovation. Although the phrase "instructional technology" may seem new to some, its roots stem back over several centuries to the origins of the "picture book as an aid to teaching as early as 1654 by Comenius" (Knapper, 1980). According to Saettler (1968), other early pioneers who contributed to "instructional technology" include Edward L. Thorndike, an American educational psychologist, who fashioned the first scientific learning theory and established empirical investigations as the basis for a science of instruction. Thorndike formulated a theory of connectionism, which was based upon the conditioning of reward or punishment, success or failure, and satisfaction or annoyance to the learner. His idea of the reflex arc, which connected the brain and neural tissue with the total behavior of the organism, ended the search by eliminating it as a separate entity. His idea of connectionism also placed the total response of the learner within his/her environment. Thus, Thorndike formulated laws of learning, which provided the basic principles leading to the technology of instruction. His studies on instructional media, the organization of instruction, individual differences, and methods of evaluation were both extensive and original.

Saettler further revealed that although Thorndike was the exemplar of empirical theorizing and investigation, his theories were rejected by many educational leaders who were attracted to John Dewey's more democratic approaches of instruction and learning, rejected Thorndike's theories. Dewey, the pragmatist and founder of the experimental school, built a system that had little basis in empirical data and whose hypotheses were not subjected to experimental investigation, despite his warning to inquire, test, and to criticize. Dewey's importance in instructional technology stems primarily from his vast influence on American education and from his analysis of thinking in reflective, problem-solving terms. Dewey was responsible for the application of pragmatism to education – the notion that education is life. In contrast to Thorndike, Dewey believed that stimulus and response were not sharply distinguished but were seen as organically related. Dewey disagreed with Thorndike's theory of connectionism and attacked Thorndike's "reflex arch concept," contending that learning involved interaction or two-way action between the learner and his environment (Saettler, p. 53). He also discussed

Dewey's experimental school. In 1896 while at the University of Chicago, Dewey decided to establish a Laboratory School for the purpose of testing his educational theories and their sociological implications. During the seven years of its existence, the Laboratory School, which advanced progressive education, became the most interesting experimental endeavor in American education. However, in 1903, Dewey declared progressive education a failure and closed the school down, realizing that it had too hastily destroyed the traditional instructional pattern without replacing it with something better. Perhaps his most enduring contribution to technology of instruction was his concept of instruction in terms of scientific method. To Dewey, all worthwhile thinking was reflection or the "active, persistent, and careful consideration of any belief or supposed form of knowledge" (Saettler, 1968, p. 323). His instructional approach resembled that of a scientific investigation in which hypotheses could be formulated and tested. Dewey believed that the primary goal of instruction was the improvement of intelligence.

Another pioneer to mention in the instruction of technology is William H. Kilpatrick, one of Dewey's students and disciple at Columbia University. After completing his doctoral degree and becoming an eminient faculty member at the Teachers College, Kilpatrick simplified and clarified Dewey's complex thinking and writing as well as added his own interpretations. He became known as "the million-dollar professor" because his estimated thirty-five thousand students had paid over a million dollars in fees to Columbia University. Kilpatrick developed the "Project Method" in an effort to present a purposeful approach to instruction. He reorganized the curriculum as a succession of projects suitable to the interest of learners. At the peak of his career, Kilpatrick's progressive slogan in the education arena became "children learn by doing," which distinctly implied Thorndike's connectionist psychology of learning, not Dewey's problem-solving methods (p. 324).

An important pioneer in nourishing a science of instruction was Maria Montessori, an Italian educator and the first female to receive a medical degree from the University of Rome. Her dominating interest in the development and welfare of children soon diverted her attention from medicine to education. By 1907, Montessori had started a state school for mentally deficient children. She trained a resident teacher for each school, selected the instruction materials, and devised techniques derived partly from Edward Seguin, a French educator, noted for his outstanding work with mentally retarded children. In 1909, Montessori became a world-renowned author and theorist. Two of the basic principles of her teaching methods included respect for the learner's individuality and encouragement of his freedom-determined not only the psychological climate and physical arrangement of the classroom but also the relationship of the teacher and learner, the instructional media, and the nature of instructional procedures (p. 325).

According to Saettler (1968), the instructional materials used by the children were self-corrective so that learners could discover their own mistakes and become progressively more independent of the teacher. Montessori emphasized the senses, individuality, and association of other learners with one another, working particularly with visual, muscular, tactile, and auditory sensations. Montessori's emphasis on "sensory learning" was based on her careful observation of mentally retarded children. Her theory, which proved to be closer to reality than the theories of her critics, was that children had a spontaneous interest in learning and that motivation was inherent in the child's interaction with his or her environment (p. 327).

Nevertheless, in 1925, Sidney Pressey noticed that his students learned at a faster pace when he used a simple machine that he devised to present and score tests without using a pencil or pen. According to Knapper (1980), after Pressey developed the first teaching machine, B. F. Skinner of Harvard University studied and applied the learning machine to his work as a way of providing a practical demonstration of operant conditioning (theory of learning) as the key to learning. As a result of Skinner's research and findings in the early 1950's, instructional technology was closely linked to psychological theories of learning (p. 20). The history of instructional technology has made significant strides over the last seventy-five years. With the profound changes and challenges facing industries, businesses, and higher education, the global society has also changed its direction with instructional technology as a new pedagogical method for students' needs and challenges in the 21st century.

For the purpose of this book, the term "instructional technology" is defined as "the theory and practice of design, development, utilization, management, and evaluation of processes and resources for learning" (Seels & Richey, 1994). The Association for Educational Communications and Technology (AECT), a well-known organization in higher education and in instructional technology, adopted this definition to justify its position. AECT representatives state that it adequately accommodates the institutionalization of Internet-based instructional processes and utilization.

Other researchers and technologists such as Tom Cutshall (1999) defined instructional technology on his homepage more specifically as the research in and application of behavioral science and learning theories and the use of a systems approach to analyze, design, develop, implement, evaluate and manage the use of technology is to assist in the solving of learning or performance problems. (p. 1)

To others, instructional technology has been defined more generally as "a process of teaching and learning with the aid of various multimedia tools" (Esther, 1998). Ely (1997) refers to it simply as "the use of such technological processes specifically for teaching and learning" (p. 1).

According to Jacobsen (1998), the integration of technology into the teaching-learning process has changed the role of students from being "passive recipients of content to becoming more active participants and partners in learning" (p. 1). Readings from selected journal articles and personal testimonies from faculty members in higher education indicate that more instructors are using technology; however, the distribution of technological innovations for teaching and learning has not been widespread, nor has it become deeply integrated into the curriculum. Although there are a growing number of faculty who are very enthusiastic about adopting instructional technology in their teaching strategies, there are still a large number of faculty who seem reluctant to adopt technology for the sake of broadening students' experiences and explorations in this concept.

Some researchers concur that students can access the same instructional resources as a fixed campus in a variety of forms, regardless of their location whether in their homes, on the campus, or in their work place (Marien, 1996; Moore, 1997). The authors contend that if instructional technology is so good, why isn't everyone using it?

Still, others claim that technology requires too much time and effort, supplies too many distractions, and yields too little value for the investment (DeSieno, 1995). Many discount instructional technology in higher education as dubious, inconsistent, and ephemeral (Winner, 1995; Shields, 1995). Incorporating instructional technology into the classroom may prove to be a timesaving tool not only for instructors but also for students as well. Both the instructor and students will be able to communicate to each other at any time outside of the classroom. Also, students can communicate with each other during other times other than in the classroom.

Everett M. Rogers's Diffusion of Innovation

The majority of the literature encompasses the diffusion of innovations theory. Rogers, an eminent scholar and well known theorist in the diffusion of innovations, specialized in the study of innovations in the early 1950's and has become a leading expert on this subject. The diffusion of innovations theory provides an approach to discussing the differences between early adopters and others. The importance of a theoretical framework is rooted in the cycle of knowledge development; whereas, observations lead to theory to classify, explain, and predict the outcome. The theory leads to questions about an individual's behavior or actions being observed. Theory-based research defines expected outcomes and the variables associated with them, and it provides a reason for expecting to find certain results. Research provided evidence for or against the theory-based expectations.

According to Rogers (1995), the history of diffusion research was traced back to the beginning of the 20th century in German, Austrian, and British schools in anthropology. Diffusion is the process by which an innovation is communicated through certain channels over time among the members of a social system. Intrigued by his research on the early pioneers of diffusion of innovations and in part because of a tremendous intellectual transfusion of academic talent from Europe, Rogers asserted, "The roots of diffusion research extend back to the European beginnings of social science with a French sociologist and judge, Gabriel Tarde" (1995, pp. 38-41). From his curiosity and research, Tarde was responsible for the S-shaped curve as indicated in Figure 1, which shows the level of adoption versus time for an innovation. Gabriel Tarde set forth a theory of imitation of how individuals were influenced by the behavior of others with whom they came in daily contact. The evolution of imitation provided a basis for communication research on the diffusion of innovations and on social learning theory.

Figure 1. The S-shaped curve is used when data are gathered from adopters asking respondents to look back in time after the innovation has diffused widely.

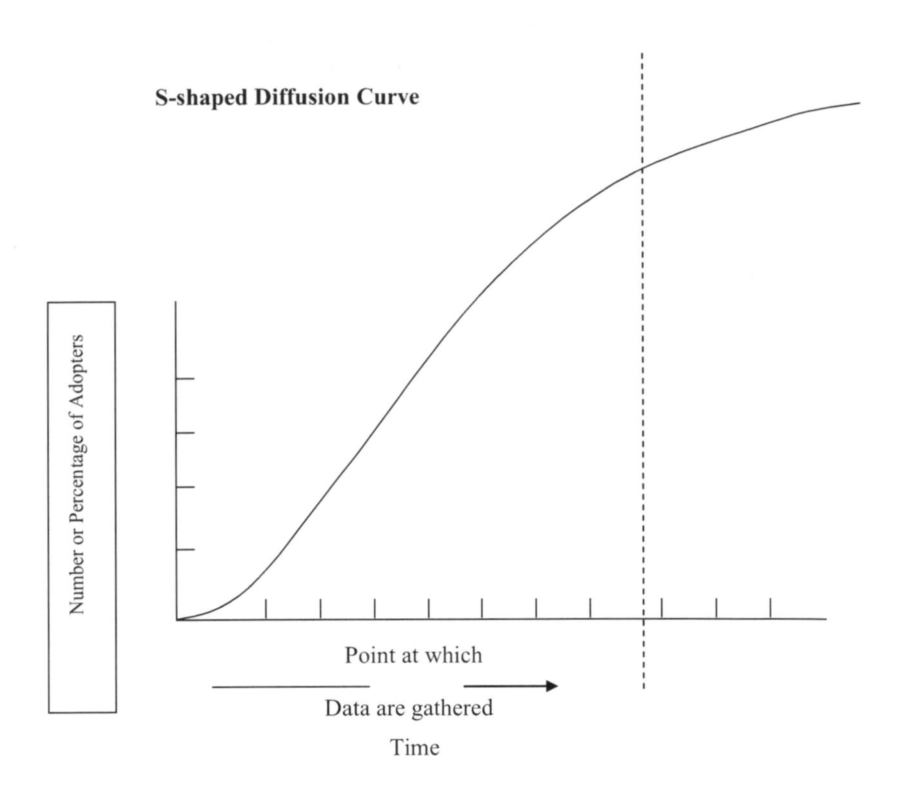

Note: From *Diffusion of Innovations* (p. 106), by E. M. Rogers, 1995, New York: The Free Press. Copyright 1995 by the Division of Simon & Schuster, Inc. Reprinted with permission from the author.

An example of how the rate of adoption might typically be represented by an s-curve states that following the period of rapid growth, the innovation's rate of adoption will gradually stabilize and eventually decline. As with all innovations, past studies have shown that new ideas typically start at the left-hand tail of the s-shaped diffusion curve, with the first adopters accepting it before it reaches the critical mass. When decisions and events take place before potential adopters reach a saturation point, the innovation itself must have a strong influence on the diffusion process.

According to Rogers (1995), the evolution of diffusion did not attract a serious audience until 1943 when sociologists Bryce Ryan and Neal Gross published a study dealing with the adoption of hybrid seed corn by Iowa farmers. This study focused on the diffusion and adoption of a new type of corn, which was adopted by almost 100 percent of the Iowa farmers over a 13-year period. The two researchers studied the rapid diffusion of hybrid corn in order to obtain lessons learned that might be applied to the diffusion of other farm innovations. Ryan and Gross's study concluded that not only did this research put diffusion of innovation on the academic map but also it made researchers realize the importance of the communication process (p. 44). Since that time, over 3,000 other studies have been conducted on the subject.

By 1960, the study of diffusion of innovations had all but ceased. Then Rogers, a young scholar and behavioral theorist, decided to carry the study of diffusion into disciplines such as communication, education, public health, marketing, general sociology, and economics. In his book, *Diffusion of Innovations (1995)*, Rogers discussed the difficulties of getting new ideas adopted, even when they had obvious advantages along with the disadvantages. As a result, Rogers noted that innovations often require many years from the time they become widely available until the time they are accepted or rejected based on likes, needs, or discontinuance of a new idea.

In terms of Rogers's diffusion of innovation model, students' acceptance of moving technology into the instructional program is influenced by certain identifiable variables such as attitudes, perceptions, and expectations. The research questions for this study were presented as a major hypothesis followed by three sub-hypotheses:

Sub-hypotheses:

1. There is a positive correlation among student's ***attitudes*** toward the infusion of technology into the instructional program and their acceptance of instructional technology.
2. There is a positive correlation among students' ***perceptions*** toward the infusion of technology into the instructional program and their acceptance of instructional technology.

3. There is a positive correlation among students' ***expectations*** toward the infusion of technology into the instructional program and their acceptance of instructional technology.

For purposes of this study, a diffusion of innovation involves individuals who evaluate an innovation that they are considering adopting, rather than on the basis of scientific research by experts. Basically, there are subjective evaluations from those members who are influenced by their social groups and serve as social models, whose innovation behavior tends to be imitated by others in their social system.

Variables Used in the Study

To clarify what is meant by attitudes, perceptions, and expectations, the following operational terms are given. One of the earliest definitions of ***attitude*** was developed by Thomas and Znaniecki (1918) as "a mental and neutral state of readiness, organized through experience, exerting a directive or dynamic influence upon the individual's response to all objects and situations with which it is related" (Simonson & Maushak, 1996, p. 985). Since that time, Zimbardo and Leippe (1991) have defined attitude as "an evaluative disposition toward some object based upon cognition, affective reactions, behavioral intentions, and past behaviors that can influence cognition, affective responses, and future intentions and behaviors" (Simonson, 1995, p. 365). For purposes of this particular study, the ***attitude*** component can be viewed as an internal factor when comparing interpersonal traits such as one's feelings or motivation toward an innovation. ***Perception*** is defined as "recognition and interpretation of sensory stimuli based chiefly on memory, insight, intuition, or knowledge gained by perception or discernment" (Simonson, 1995). In a similar meaning, perception is defined as "the process of organizing and interpreting sensory information, enabling us to recognize meaningful objects and events" (Myers, 1998, p. 148). For this study, perception can be viewed as an external factor, which concerns one's view, understanding, belief, or reaction to an innovation. The third variable to be considered in this study is expectations. According to Simonson (1995), ***expectation*** is defined as "what one anticipates or looks forward to as a result of the previous experience" (p. 171). In this study, expectations were viewed as one's belief, understanding, or conception about an innovation. Therefore, within the context of Rogers's diffusion of innovations concept, the researcher investigated the attitudes, perceptions, and expectations of students toward instructional technology as the three major components in the study.

Significance of the Study

This investigation should expand the present knowledge of the purported relationship between what students are learning in higher education, how they are adopting to the change in technological pedagogy, and what their specific needs are in terms of effecting a change in their learning environment. Therefore, this study of diffusion was important to research for several reasons:

First, administrators and faculty of public four-year institutions must realize the validity of technological instruction as a learning tool in the acquisition of knowledge by students. Research has shown that student achievement is the paramount objective of most classroom activities. Therefore, it was important to recognize the need for establishing attitudinal changes among students and for planning learning activities designed to facilitate attitudinal outcomes in the new learning environment.

Second, the most powerful rationale for the need to promote attitudinal change in students was to demonstrate a direct relationship among students' attitudes, perceptions, and expectations toward technology and their adoption to changes in their learning environment, associated with the introduction to instructional technology. Although there has been considerable research investigating this relationship, there was little empirical evidence to establish the validity of the linkage among the three constructs of attitudes, perceptions, and expectations.

Third, in some instances, influencing the attitudes of students has not always been desirable; therefore, administrators and instructors should be aware of which techniques affect students' attitudes. In this way, possible bias can be recognized and eliminated. As a result, students' behavioral characteristics such as attitudes, perceptions, and expectations toward instructional technology can be further investigated to reveal how instructional technology can impact student learning.

It has become increasingly apparent to those involved in educational technology research that the consequences of computer integrated instruction are the results of positive attitudinal change in students. The findings of this study will contribute to a better understanding of the relationship among students' behavioral characteristics of attitudes, perceptions, and expectations toward instructional technology and their learning environment associated with the introduction to instructional technology. Also, it will add to the growing body of knowledge about the diffusion of innovations and its impact on higher education and social learning.

Due to time constraints, the investigator was assisted by collaborating professors and students at two four-year public institutions. However, the investigator was able to limit the study to responses on a 40-item survey, which was administered to approximately 171 undergraduate students. As a result, a correlation study was conducted among students who had some or no prior experience with instructional technology courses.

Assumptions

Data gathering was accomplished without any disruptions to the students or professors in the classroom settings. Based on prior research by Carr (1999) and Ponton (1999), a 10-point Likert scale was developed to assess adult autonomous learning toward acceptance of technology and was considered a valid instrument to use. The investigator also used a similar scale to assess students' attitudes, perceptions, and expectations toward instructional technology. The instrument used in this study was the Students' Attitudes, Perceptions, and Expectations Toward Technology (SAPETT) survey, which was created by the investigator and was administered to students with no added complexities to the classroom environment. The investigator also conducted a pilot-test with 15 students who were given the SAPETT survey, which indicated that it was a reliable instrument to use for this study. The assumption that the students' behavioral characteristics in a correlation study would not differ except that their behavioral characteristics were influenced by previous experience with instructional technology, socioeconomic status, and perhaps gender.

CHAPTER 2

An Introduction to Instructional Technology

This chapter describes literature relevant to the research purposes of this study. It is organized into five sections: (1) an introduction to instructional technology, (2) the theory of diffusion of innovations, (3) students' attitudes toward technology affected by the introduction of instructional technology, (4) students' perceptions toward technology affected by the introduction of instructional technology, and (5) students' expectations toward the learning process affected by the introduction of instructional technology. At the end of each section, the relevance of the literature to the research reported in this study is discussed.

Technology has advanced rapidly over the past several years, and there have been numerous published studies investigating its impact on educational needs in higher education. The "change agents" or those who have been forced to adopt the pedagogical use of instructional technology as an innovative method for teaching and learning are still going through an extensive process called the evolution of technology. For example, a report done by Pea and Soloway (1987) for the U. S. Congress Office of Technology Assessment stated, "Technology might be the factor which would help bridge the over-widening gaps between schools and society" (pp. 33-34). Of course, society as a whole and higher education professionals did not accept the innovation as quickly as was expected. The use of technology, with the help of the electronic calculator and the computer, was accepted by some innovators and early adopters, stalled by the critical mass (the early and late majority population), and rejected by the laggards (those who accepted the old way of doing business with pencil and paper).

While some researchers concluded that technology was the beginning of new ideas, Kerr (1991) predicted that some educators would plunge into technology with no regard for understanding the consequences. He insisted that from the inception of an innovative concept, there were those who were eager to start on a new quest; however, there were some who did not want to start anything new. Such endeavors have been the case for many educators, but even those willing to engage into technology have often times not had opportunities or access to equipment and training. Although most researchers agree that students in higher education must be ready for the challenges of global competition, they also realize that most instructors are ill-equipped or not properly trained to integrate technology-mediated strategies as a new method of teaching. Students must have the capabilities of expanding distance education through new technologies, strategize for learning throughout a lifetime, and increase their chances for success with basic competencies in higher order problem solving.

Some researchers have argued that this learning revolution through instructional technology has come about relatively slow. The slow pace of change and irrelevant curricula in the face of increasing technology and an emerging global economy have put higher education behind the line to be replaced by business enterprises (Perelman, 1993; Burke, 1994; and Denning, 1996). However, some educational leaders are well aware of the competitive environment students must confront and have recognized the relevance of technology to the knowledge, skills, and wisdom that students need to navigate through a technologically-based society (Ehrmann, 1995). The new trend of technology in higher education will incorporate solutions aimed at improving learning productivity by reducing labor intensity and providing new ways to deliver education and better services to students while enhancing the quality of instruction (Baker & Gloster, 1994).

In today's society, higher education is faced with a plethora of challenges. Among those challenges, faculty are slowly being forced to adopt and deliver quality instruction by incorporating technology. Faculty members are struggling to keep up with the innovative channels of instructional technology as a learning tool for college students. This challenge is not unique to the twenty-first century, however.

Theoretical Framework

A theoretical framework for analyzing the behavioral characteristics of undergraduate students was based on the theory of the diffusion of innovations advanced by Everett Rogers. This analysis was drawn from his book, *Diffusion of Innovations* (1995), which used the structural, human resource, and symbolic frames to explicate the diffusion theory. As Rogers noted, the advantage of an innovation, "A technological innovation usually has at least some degree of benefit for its potential adopters. This advantage is not always very clear-cut, at least not to the intended adopters. They are seldom certain that an innovation represents a superior alternative to the previous practice that it might replace" (1995, p. 13). There was a positive connection between the theory of the diffusion of innovations by Rogers and the specific innovation of interest in this investigation. Rogers (1995) defined diffusion as "the process by which an innovation is communicated through certain channels over time among members of a social system" (p. 5). The key elements are the innovation, communication channels, time, and the social system. In his theory of diffusion of innovations, Rogers described the components of movement toward an innovation as a time line process into the curriculum of any study until diffusion is reached. Diffusion consists of four main elements:

Innovation – any item, thought, or process that is viewed to be new by the individual.

Communication channels – the process of the new idea traveling from one person to another or from one channel to the individual(s).

Time – a period it takes for the group to adopt an innovation as well as the rate of adoption for the individual(s).

Social system – a group of individuals working together to complete a specific goal: adoption.

The first main element of the diffusion process is innovation. Innovation is defined as "an idea, practice, or object that is perceived as new by an individual or other unit of adoption" (Rogers, p. 11). In essence, innovation denotes an idea, thought, or process that is new to individuals in a social system.

The second main element of the diffusion process is "communication channels." Communication is the process by which participants create and share information with one another in order to reach a mutual understanding. The "communication channel" is the medium by which messages get from one individual to another. In a comparative analysis of interpersonal communication and mass communication channels, interpersonal communication channels are more effective in forming and changing attitudes toward a new idea that may lead to influencing an individual's decision to adopt or reject an innovation.

The third element of the innovation process, according to Rogers, is time. The time dimension affects diffusion in three ways. First, time is a critical part of the innovation-decision process. Second, time is involved in diffusion through the innovativeness of an individual or other unit of adoption. Innovativeness is the degree to which an individual or other unit of a system is relatively earlier in adopting new ideas than other members of a social system. Third, time is involved in the diffusion of the rate of adoption. The rate of adoption is the relative speed with which an innovation is adopted by members of a social system. The rate of adoption is usually measured as the number of members of the system that adopt the innovation in a given time period.

The fourth element in the diffusion of new ideas is the social system, which is defined as a set of interrelated units that are engaged in joint problem solving to accomplish a common goal. The members or units of a social system may be students, individuals, informal groups, organizations, or other subsystems. The social system constitutes a boundary within which an innovation diffuses. Together, these four critical elements constitute an innovation.

A Model of the Five Steps of the Innovation-Decision Process

Rogers (1995) defined an innovation as "an idea, practice, or object that is perceived as new by the individual," and he defined diffusion as "the process by which an innovation makes its way through a social system" (p. 11). The characteristics of innovations, as perceived by individuals, tend to influence their rate of adoption and are associated with the persuasion stage of the innovation-decision process (see Figure 2). Rogers (1995) listed five perceived characteristics of an innovation:

Relative advantage - the degree to which an innovation is perceived as better than that which it supersedes.

Compatibility - the degree to which an innovation is consistent with the existing values, past experience, and needs of the potential adopter.

Complexity - the degree to which an innovation is perceived as difficult to understand and use.

Trialability - refers to whether an innovation may be experimented with on a limited basis.

Observability - the degree to which the results of an innovation are visible to others. (pp. 212-244)

The innovation-decision process is the process through which an individual or other decision-making unit passes from first knowledge of an innovation to forming an attitude toward the innovation, to a decision to adopt or reject, to implementation and use of the new idea, and to confirmation of this decision (Rogers, 1995, p. 161). Rogers's model of the Innovation-Decision Process, there are four conditions (previous practice, felt needs or problems, innovativeness, and the norms of the social system) which precede the communication channels. Each time an individual accepts or rejects an innovation this five stage process is implemented as depicted in Figure 2.

Figure 2. A Model of the Five Stages in the Innovation-Decision Process.

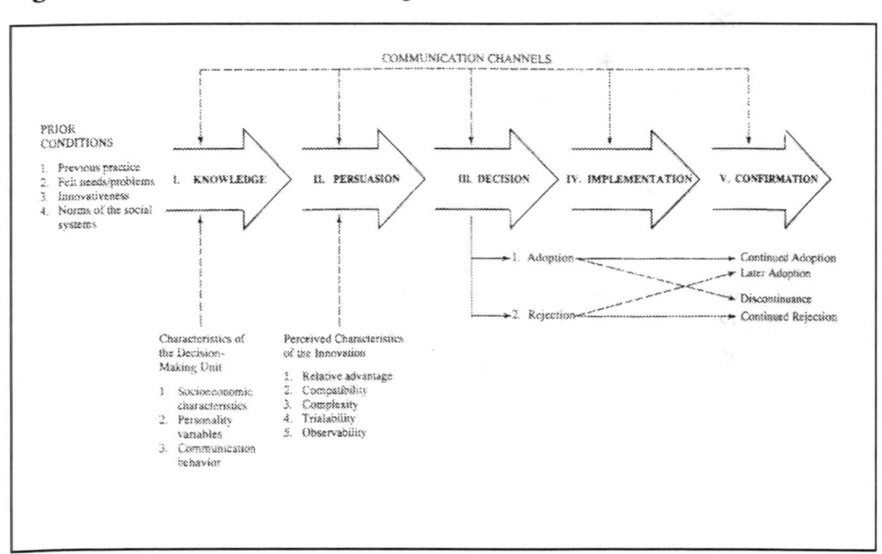

Note: From Diffusion of Innovations (p. 163), by E. M. Rogers, 1995, New York: The Free Press. Copyright 1995 by the Division of Simon & Schuster, Inc. Reprinted with permission from the author

Next, a series of actions and choices over time through which an individual evaluates a new idea and decides whether or not to incorporate the innovation into an ongoing practice. The concept of the communication channels consists of knowledge, persuasion, decision, implementation, and confirmation. Rogers (1995) gave a detailed conceptualization of the five stages of the innovation-decision process as viewed among individuals to include:

1. ***Knowledge*** occurs when individuals are exposed to an innovation's existence and gains some understanding of how that innovation functions.
2. ***Persuasion*** occurs when individuals form favorable or unfavorable attitudes toward the innovation.
3. ***Decision*** occurs when individuals engage in activities that lead to a choice to adopt or reject the innovation.
4. ***Implementation*** occurs when individuals put an innovation into use.
5. ***Confirmation*** occurs when individuals seek reinforcement of an innovation-decision already made, or they may reverse a previous decision to adopt or reject the innovation if exposed to conflicting messages about the innovation.

A recent search of the Educational Resources Information Center (ERIC) for the years 1997 through 1999 using the subject index terms, "instructional technology" and "students' attitudes toward technology," resulted in several studies on technology in higher education. One study revealed that although higher education researchers were integrating technologies into their teaching strategies for adult learners, they were not considering how technology could be used to support and expand adult learning (Imel, 1998). In her study, Imel concluded that even though technology was being integrated into the curriculum as a delivery mechanism, as a complement to instruction, and as an instructional tool, it presented some limitations to the students as a means for learning. However, she concluded that technology could be structured to capitalize on the individual characteristics of adult learners and to minimize the limitations of the instructors or the multimedia itself. Imel also argued, "The primary role should be to ensure that the focus is on learning and not on the technology" (p. 4).

Another study revealed that if higher education wanted to advocate change in the way students were learning, educators were urged to place learning first and redesign their current learning environment inherited from the past. The main thrust of the article emphasized the paradigm shift from the instructional mode to students learning with technological advances (Schuyler, 1998).

Van Dusen (1997) discussed the impact of instructional technology in higher education and saw the virtual campus as a metaphor for the electronic teaching, learning, and research environment, which was created by the convergence of powerful new information and instructional technologies. He explained that colleges and universities were serving a more heterogeneous clientele with diverse educational backgrounds and needs. Therefore, instructors need to recommit to creating an ideal learning environment for students, employ new technologies to address their needs, and change from the traditional model of the classroom as a teacher-centered to that of a student-centered classroom. Van Dusen believed that technology would be adopted if courses were modified to accommodate the use of instructional technology in the curriculum and if faculty would change from being content providers to facilitators. At the same time, students would be required to adopt a more proactive stance in their own education. In addition, Van Dusen asserted that colleges and universities must become learning organizations that foster originality and innovation. In order to begin the process of integrating technology into higher education, Van Dusen recommended these seven steps:

1. Create a venue where key stakeholders can analyze major technology issues and purchases;
2. Assert the value of technology-based learning from a variety of research perspectives;
3. Establish quality standards for certificate and degree programs;
4. Avoid pitting traditionalists against technology enthusiasts;
5. Make collaboration and cooperation, not reengineering and restructuring, the new institutional buzzwords;
6. Retain a strong commitment to adequate library staffing and funding; and
7. Prepare for success by creating the necessary support structures. (p. 37)

Once these steps are put into place, Van Dusen argued that educators and administrators would see the pressures lessen and harvest the benefits of their students' success as productive citizens.

At present, the use of technology is growing and simultaneously changing the learning process, the structure of knowledge, and the nature of instruction, including curriculum development and assessment. Through his survey on *Campus Computing 1995*, Kenneth Green (1996) illustrated that major gains have been made in the proportion of colleges and universities that use information technologies as instructional resources. In another study conducted by Elaine El-Khawas and Linda Knopp (1996), senior administrators at some 506 colleges and universities were asked to complete survey questionnaires, and some 403 administrators responded to questions about various issues facing their institutions in 1996. The results of the study revealed that half of all the institutions increased their attention to teaching and learning as the most significant program changes in the last decade. Most respondents indicated an increase in the use of technology; however, El-Khawas and Knopp asserted, "Only 29 percent revealed an ability to keep up with the latest technological advances" (*Campus Trends*, 1996, p. 2). Other findings in the study related to widespread curricular changes among the colleges and universities: 98 percent of the administrators cited instruction using computer technology as a major change. Approximately 72 percent of public institutions offered courses by interactive television or by other electronic means, and one-quarter had taken steps to offer courses using the Internet (p. 23). El-Khawas and Knopp concluded that due to the increased changes and pressures on higher education, most colleges and universities found it a difficult task to keep up with the changes in computer and instructional technologies. They added, "The challenge ahead will be a necessity, not a luxury – posing serious problems of funding and capacity-building for all colleges and universities, whatever their current resources" (p. 28). Likewise, Green's national survey in *Campus Computing* showed a dramatic increase with computer technology but also indicated that colleges and universities were having difficulty maintaining a comfortable pace with computer technology (*Campus Computing* 1996; 1997; 1998; and 1999).

The Critical Mass

While research on the positive results of faculty development with the use of technology in learning continues to grow, few studies have been conducted on the practice of integrating technology into the classroom for the benefit of students. Some researchers have expressed their concerns about how post-secondary education institutions are employing distance learning and technology-based instruction (Cartwright, 1998; Gilbert and Ehrmann, 1995). Moreover, research is limited on specific ways that technological techniques are being integrated into the classroom for students based on need, preference, social status, and lifelong learning.

A 1999 study on instructional technology was conducted at Middle Tennessee State University (MTSU) to examine the effectiveness of its use. Results provided valuable information that helped to determine students' feelings, beliefs, and satisfaction with the course outcomes related to improving technological resources for the University's faculty and students. A survey was administered to and returned by eight percent of the student population of 18,000 students. Results showed that the use of instructional technology revealed positive effects from the students, increased student interest and satisfaction, and was formed to facilitate an important role in their learning activities (Draude & Brace, 1999).

In another technology study, Twigg (2000) found that the Universities of Idaho and Alabama, and Virginia Tech were leveraging the power of information technology to establish individualized learning environments suitable to their students' needs. All three universities reduced the cost per student, while increasing the quality of the learning experience for students (*The Pew Learning & Technology Newsletter*, December 2000). Subsequently, Rogers explained that diffusion was just a phase of the larger sequence through which an innovation goes from the time the decision to begin research on a recognized problem is made to the consequences of that innovation (p. 131). To be specific, before the diffusion process begins, several preliminary steps must occur: a problem is perceived, funding decisions about research and development activities lead to research work, the innovation is invented, then developed and commercialized, a decision to diffuse an idea is made, a transfer of the innovation is made to a diffusion agency, and its communication to an audience of potential adopters is reached. Then, the first adoption occurs and the diffusion process begins. To complete the package, a pre-diffusion series of activities and decisions are derived from the innovation-decision process. The innovation-decision process consists of all the decisions and activities, the impact that occurs from recognition of a need or a problem, through research, development and commercialization of an innovation, through diffusion and adoption of the innovation by users, and through its consequences.

As a result, the innovation-decision process is driven by the exchange of technical information in the face of a high degree of uncertainty. Rogers (1995) noted that one factor influencing the adoption of an innovation is complexity. Complexity applies to an innovation viewed as complicated, difficult to understand or to use; such an innovation is less likely to diffuse. Trialability and observability are two factors that provide the capacity to experiment with and see the results of the innovation. Innovations that have greater relative advantage, compatibility, trialability, observability and less complexity will be adopted more rapidly than other innovations. Compatibility of the innovation is measured by comparing existing values, past experiences, and the needs of the adopters. Rogers stated that the adoption of an incompatible innovation often requires the prior adoption of a new value system, which is a relatively slow process (p. 16). Within the social construction of technology, technological determinism is the belief that technology causes change in a society. Although technology is a product of a society, it is influenced by the norms and values of the social system. Consequently, some individuals evaluate an innovation on the basis of scientific research; however, others will base their decisions on the subjective evaluations of friends or peers. On the other hand, mass media channels are more effective in creating knowledge of innovations that reach out to a larger group, which Rogers called "the critical mass" (p. 313).

Geoghegan (1998) maintained that the "critical mass" or mainstream audience in higher education had not yet reached full saturation in the use of instructional technology. In order for that to happen, he suggested introducing Rogers' diffusion of innovation attributes: relative advantage, compatibility, complexity, trialability, and observability to a critical mass of faculty and students and to provide technical training and support. As an alternative strategy, Geoghegan advised institutions to consider "increasing students' tuition fees by 10 – 15% to accommodate students with access to or ownership of desktop computers or to provide technical support systems" (p. 148).

To bridge the gap between faculty, students, and technology, Geoghegan asserted that if faculty are to use instructional technology effectively to improve their teaching methods, they must be assured technical support and rewarded for their efforts. As demonstrated in the rate of adoption for an interactive innovation, Geoghegan concluded, "the 85% who are the majority, constitute the mainstream or the "critical mass" (see Figure 3), and their successors are the ones who hold the key to bringing about improvements to teaching and learning" (p. 149).

The critical mass occurs at the point where individuals have adopted an innovation so that its rate of adoption becomes self-sustaining. Although Geoghegan made a strong appeal in favor of faculty, he may have overlooked two critical factors concerning all institutions - where will the money come from, and how can reluctant faculty be encouraged to find the time and patience to be properly trained?

Figure 3. The Critical Mass denotes the rate of adoption for a usual innovation and for an interactive innovation.

The Critical Mass

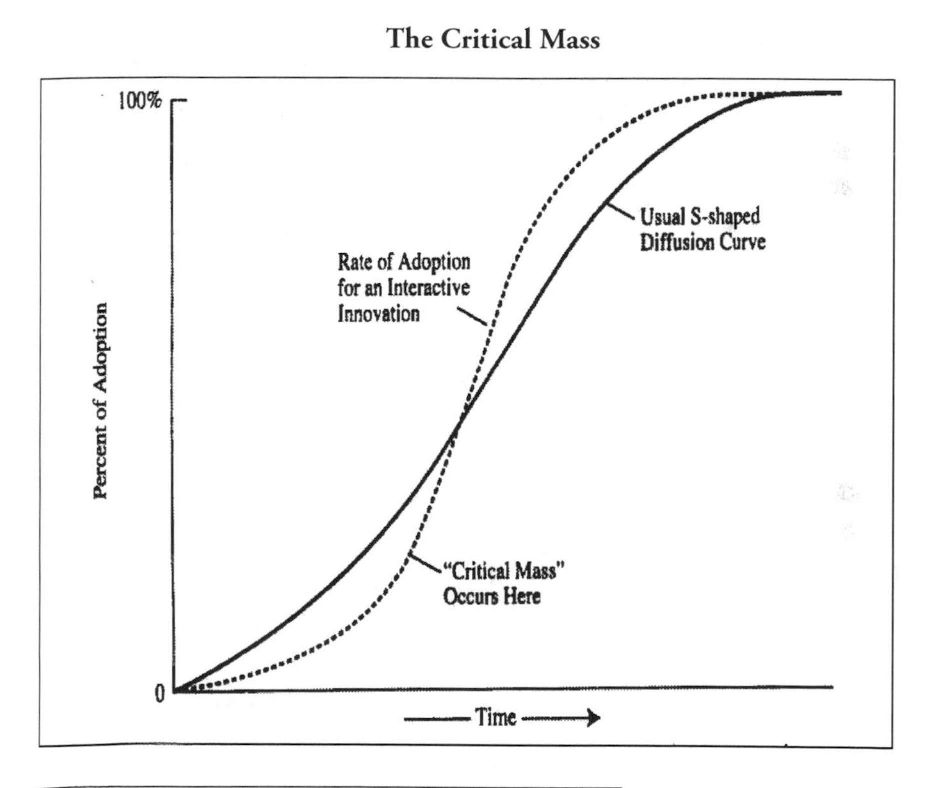

Note: From Diffusion of Innovations (p. 314), by E. M. Rogers, 1995, New York: The Free Press. Copyright 1995 by the Division of Simon & Schuster, Inc. Reprinted with permission from the author.

Adopter Categories

In reference to the individual innovativeness theory, Rogers (1995) stated that individuals who are predisposed to being innovative will adopt an innovation earlier than those who are less predisposed. Rogers's bell-shaped distribution curve is broken down into five distinct categories. On one extreme of the distribution curve are the innovators (2.5%). They are the risk takers and pioneers who adopt an innovation very early in the diffusion process. The next to adopt an innovation are the early adopters (13.5%), who are included in the area between the mean, minus two standard deviations. Then there are the early majority (34%), who are included in the area between the mean data of adoption and the mean, minus one standard deviation. The late majority (34%) are the ones who fit between the mean and one standard deviation to the right of the mean. At the other extreme are the laggards (16%), who resist adopting an innovation until rather late in the diffusion process. Figure 4 shows the bell-shaped distribution curve of Individual Innovativeness and the percentage of potential adopters theorized to fall into each category.

Figure 4. Adopter categorization on the basis of innovativeness.

The Bell-Shaped Distribution Curve

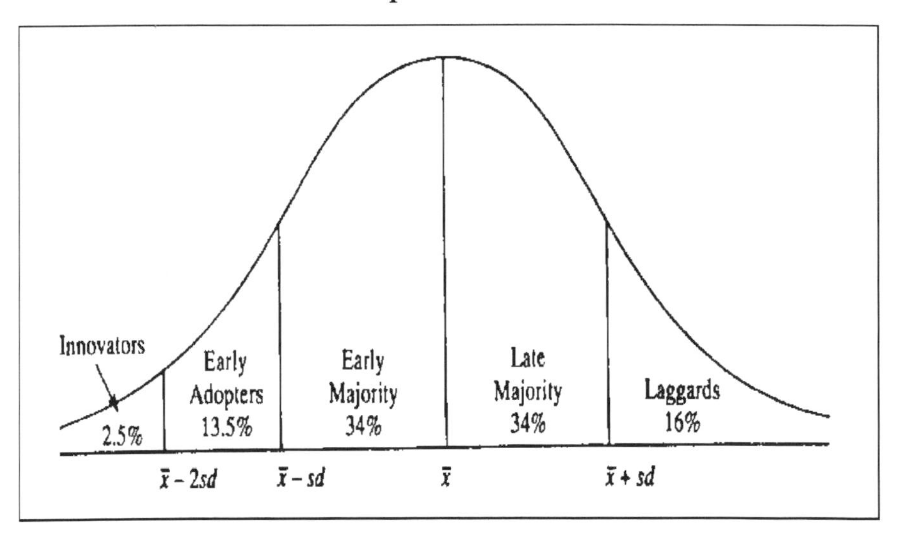

Note: From *Diffusion of Innovations* (p. 262), by E. M. Rogers, 1995, New York: The Free Press. Copyright 1995 by the Division of Simon & Schuster, Inc. Reprinted with permission from the author.

Rogers (1995), Geoghegan (1994), and Moore (1991) recognized five distinct categories of adopters distributed along a continuum closely resembling a bell-shaped curve (as shown in Figure 4). The innovative decision process, as measured by the time an individual adopts an innovation, is continuous. The innovativeness variable is divided into five adopter categories by laying off standard deviations from the average time of adoption (x). These categories of adopters are as follows:

Innovators: those 2.5 percent of the population that are characterized as techies or experimentalists who seize any new technology as soon as it starts. Innovators are in the small percentage that takes the risk with no reservations or hesitations.

Early adopters: the 13.5 percent of the adopter population who are considered to be the visionaries in applying the instructional methods of learning effective skills. They are also risk-takers, usually self-sufficient in technical skills and are broadly connected with the academic community.

Early majority: those adopters who comprise 34 percent of the adopter population are individuals who make up the first half of what is called the mainstream. This group relies on a wait-and-see attitude and seek value in an innovation before adopting it.

Late majority: those adopters who comprise another 34 percent of the adopter population and are categorized as the conservative, skeptical portion of the mainstream. They accept technology late in the process once it has become well established and easy to use.

Laggards: this group, which is comprised of 16 percent, is unlikely to employ instructional technology at all. Sometimes this group may be the last to adopt or not adopt at all because they are antagonistic toward others who have adopted the innovation.

The Chasm

After researching Rogers's work, Moore (1991) noted that in the area of high technology innovations, the differences between the early adopter population and what he called the mainstream are so great as to create a metaphorical chasm between them (see Figure 5). Moore argued, "One high technology innovation after another has floundered in the gulf between the early adopters and the mainstream" (p. 135). The chasm, in the adoption process, occurs at the point where an innovation would normally pass from the early adopters to the mainstream adopters after it has been accepted by about 15 percent of the total population. Similar to what Rogers discovered in his research, Moore found the critical mass or main-stream population to be the most popular group to convince others to try an innovation.

Figure 5. Geoffrey Moore's Chasm

The Chasm

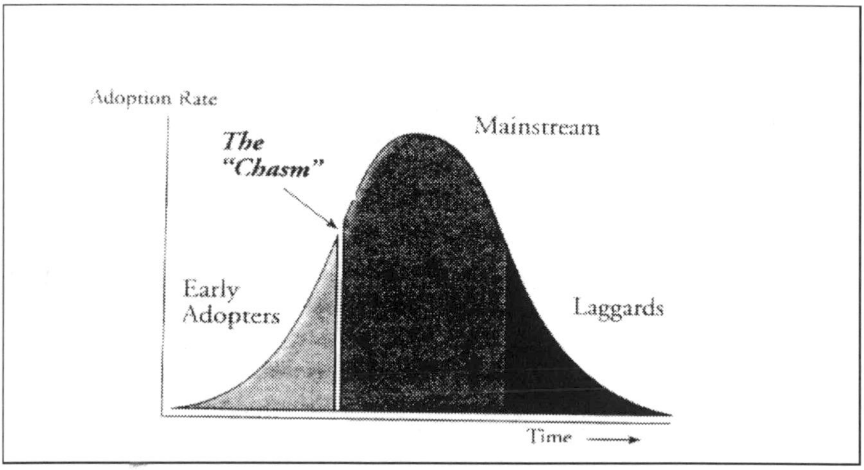

Note: From *The Chasm* by G. M. Moore. From W. H. Geoghegan (1998), Instructional Technology and the Mainstream: The Risks of Success. In D. Oblinger & S. Rush (Eds.), pp. 131-150, *The Future Compatible Campus*, Boston, MA: Anker Publishing.

Geoghegan (1998) summarized the significance of these categories and their characteristics by stating, "One of the principal reasons for the failure of instructional technology is the inability to cross the gap that separates early adopters from the mainstream. We seem to have assumed a sort of homogeneity of individual willingness to experiment with and use instructional technology, thereby ruling out the possibility of recognizing qualitatively distinct subgroups with different attitudes toward technology and its use in instruction" (*The Compatible Campus*, p. 145).

Students' Attitudes toward Instructional Technology

The first component of this study dealt with students' attitudes toward technology and how they were affected by the introduction to instructional technology. According to Kershaw (1996), instructors must change or face devastating consequences as he argues, "For those charged with implementing significant change in post-secondary institutions, it is useful to remember that, while the particularities of the change process may be terra incognito for the institution, the change process itself is well understood: Change or atrophy" (*Educational Technology*, p. 44). He also asserted that institutions should look at preparing their students to be effective learners through the use of instructional technology. Since the diffusion of instructional technology has become a part of the pedagogical method, instructors in higher education are slowly becoming apprised of its significance and the urgency to accept it in the change process. Students who come to colleges and universities are already familiar with technology to some degree. Some students are more knowledgeable than their instructors are about the use of computer technology. As Kershaw stated it bluntly, "change or atrophy," those in higher education must adhere to the paradigm shift or cease to exist.

Simonson (1995) explained that educators have long been concerned about students' attitudes toward learning. He asserted that although students' attitudes have not been linked scientifically to their achievement, they have been considered a vital element of effective instruction. Hence, instructional technology may be viewed as another fancy expression for computer use, it is considered a complex and integrated process that involves people, procedures, ideas, devices, and organizations. In addition, instructional technology can be used for analyzing and devising problems, implementing new ideas, evaluating those ideas, and managing solutions to problems considered as a learning mechanism (p. 373).

Students tend to embrace ideas that relate to their interests, needs, and existing attitudes. They consciously or unconsciously avoid messages that conflict with their predispositions. This tendency is called "selective exposure," attending to communication messages that are consistent with one's existing attitude and belief. At other times, students may exhibit "selective perceptions," that is, they may interpret communication messages in terms of their existing attitudes and beliefs (p. 164). Rogers (1995) asserted that an individual's decision to adopt an innovation is not an instant act. Rather, it is a process that occurs over time, consisting of a series of actions and decisions. It has been mentioned previously that the innovation-decision process is "the process through which an individual passes from first knowledge of an innovation, to forming an attitude toward the innovation, to a decision to adopt or reject, to implementation of the new idea, and to confirmation of this decision" (p. 163).

This theory has been cited numerous times in the instructional technology literature about which Boser, Palmer, and Daugherty (1998) wrote, "Although research on student attitudes in technology education has been used to assess students' attitudes prior to curriculum development, a standardized attitude measure such as the PATT-USA has not been used to assess changes in attitude as the result of a treatment such as participation in a technology education program" (p. 3). It is logical that students who have a positive experience in a technology education program will develop a positive attitude toward technology and the pursuit of technological careers and would therefore be more interested in studying about technology. In all instructional approaches, students' beliefs that technology is difficult were significantly reduced through participation in technological activities.

Similar research on attitudes has been popular in many disciplines for more than 65 years. This construct is considered more central to social psychology than to any other academic area. Educators have been interested in attitudes because of its impact on learning, and while attitudes have not been linked to achievement, it has been long considered an important component of education for learning practices. For the purpose of this study, the definition given by Simonson and Maushak (1996) will be used. They defined attitudes as "learned predispositions to respond," which provides direction to subsequent actions. In addition, Zimbardo and Leippe (1991) have defined attitude as an evaluative disposition toward some object based upon cognitive and affective reactions, behavioral intentions, and past behaviors that can influence cognitive and affective responses and future intentions and behaviors (p. 985). These authors also stated that since attitude is acquired, it can be changed with some predictably of outcomes. As Simonson and Maushak argued, "Attitudes are related to how people perceive the situations in which they find themselves" (p. 985). At the same time, attitudes of individuals can vary in direction, in degree, and in intensity.

Simonson and Maushak (1996) postulated that attitude positions are composed of four components: (1) affective responses, which entail a person's liking of or emotional response to some situation, object, or person; (2) cognition, which is conceptualized as a person's factual knowledge of the situation, object, another person, or how much the person knows about a topic, including the individual person; (3) behavioral component of an attitude involving the person's overt behavior directed toward a situation, object, or person; and (4) behavioral intention, which involves the person's plans to perform in a certain way, even if the plans are never acted upon (p. 987). Therefore, the four components of attitude form an attitude system, which is interrelated to produce an organizing framework or mental representation of the attitude construct. Behavioral research supports the idea that actions led to the formation of a cognitive scheme, which led to the creation of attitude position of a person's observed behavior (Simonson & Maushak, 1996). Consequently, attitudes help form cognitive relationships, which in turn predispose certain behaviors.

Despite the continual efforts in educational change, less than 10 percent of the reported research on innovation has come from education. Due to the changing settings in secondary education, the vast amount of adoption pertaining to instructional technology has come from secondary or elementary education, not higher education. However, there has been a large body of research concerning the introduction and diffusion of innovations in organizations. As Blackwell, Engle, and Miniard (1995) have found, "Over 3,000 studies and discussions of diffusion processes have been published in at least 12 identifiable disciplines" (p. 728).

Although technology has greatly permeated our society and culture, the influence on daily educational practice in higher education is limited (Denk, Martin & Sarangarm, 1993). The complaint is not that technology has failed to enhance students' learning when applied, but so few faculty members have taken the time to integrate it into the instructional process (Gilbert, 1995; Green, 1997). Furthermore, Geoghegan (1994) revealed that instructional technologies have failed to penetrate the curriculum, despite many innovative models for the use of technology to enhance teaching and learning (p. 6). According to Ehrmann (1995), once faculty obtain the technology, they will automatically change their teaching tactics and course materials to take advantage of its capabilities (p. 24). However, what Ehrmann predicted has not yet occurred in higher education. The literature suggested that better training for faculty and efforts to increase their understanding of the benefits of using instructional technology as a tool could enhance their learning and teaching process. Kearsley (1993) conducted a survey of interactive multimedia projects underway in universities and found that the major issues concerning their faculty and staff interested in interactive multimedia were its effectiveness, teaching strategies, system selection design alternatives, and development time costs. Results indicated that multimedia were effective learning or teaching tools but only if changes were made to traditional learning and teaching strategies. Another example, *The Flashlight Project*, which was designed by Ehrmann (1997), is a model for evaluating the impact of informational technology on teaching and learning. The project is a means of assessing reasons for using technology education and allows educators to work together to improve teaching and learning with technology. Further research indicated that Ehrmann was very concerned about what students were learning. In his proposed question, Ehrmann suggested that there was a missing link between students and instructors: "Are students being taught the right stuff?" He argued that instructors who use instructional technology as a pedagogical method should not use it just to show and tell. Ehrmann suggested that they focus on the "knowledge, skills, and wisdom that graduates need" in order to perform well in their new working environment after graduating from college (*Change*, p. 23).

Research in the literature included several empirical studies. One practical technique for instruction using technology was based on the concept of "anchored instruction." Anchored instruction, as described by the Vanderbilt Cognition and Technology Group (1990), uses technology to provide a realistic situation for learning. While the Vanderbilt Group's studies concentrated on the cognitive consequences of anchored instruction, there was ample anecdotal evidence to support that anchored instruction also influenced attitudes. Diamond and Simonson (1988) studied filmmakers who produced persuasive films. They found that the participants believed that the presentation of authentic situations in their films were critical to the success of their persuasive messages more so than were informational films and videotapes. The assumption was that positive predispositions, developed during participation in authentic situations, oriented students to actively pursue additional learning (p. 878). Simonson and Maushak (1996) found in previous studies that if information is presented authentically and intelligently, it is likely that it will be persuasive and will be received favorably (p. 987).

Further research of the literature indicated that learners who were actively involved in what was perceived as a real event were more likely to react in an attitudinally positive way to the situation and to instruction. Johnson (1989) found that the use of discussion questions followed by a mediated situation resulted in significant attitude changes toward careers with regard to learners' confidence in their ability to be successful (p. 584). In addition, a survey of university students in Britain produced data, which indicated that attitudes toward technology were related to past scientific experiences, with no evidence to support any across-gender effect associated with attitudes (Fife-Schaw, 1987). Other research was conducted on the form of language associated with technical change, supported by the possibility of gender differences of male-oriented language in technology fields, was found to alienate females and prevented them from participating in these fields (Wilson, 1992). Another study investigated gender differences in technology used by faculty at a midwestern university. Results showed that male faculty rated their knowledge and experiences with some innovative technologies higher than did female faculty (Spotts & Bowman, 1995).

Students' Perceptions toward Instructional Technology

The second component is students' perceptions toward technology, as affected by the introduction to instructional technology. The term "perceptions" is intended to mean how one assigns different meanings to what one sees as the logic from a different perspective. To put it another way, perception can also mean a process of discerning information and an understanding it through interpretation or association. The perception of the participants of instructional technology became a guide to interpretation. Although early studies in this category reported the effects of training or the impact of technology in the workplace, several studies focused on the stages of change. There seemed to be little evidence of empirical studies pertaining to the perceptions of adopters to technology in higher education. Holloway (1996) concluded that private commercial marketing research corporations were targeting the education area as a "for profit" market. In his view, "Marketing research about consumer behavior is clearly focused on the same variable of characteristics of innovations and adopters as education. These perceptions are based on interpersonal utility comparisons, not on the economic value of technology" (1109).

According to Fitzelle and Trochim (1996), a study was conducted at Cornell University with a group of undergraduate students enrolled in a research methods class. The students were asked to complete a 20-item survey using a Likert-type scale to rank the qualitative questions. The results showed that students thought that the web site significantly enhanced their learning of course content. The authors concluded students' perceptions of performance in the course were predicted by variables of enjoyment and control of the learning pace.

Barbara Seels (1997), a well-known advocate of instructional technology, concluded that the theory of instructional technology is based on extensive relationships among variables in order for a theory to describe only a limited range of behavioral characteristics. She stated, "Theory building using taxonomic classification is important because without taxonomic structures, it will be impossible to progress towards theoretical systems. Taxonomies affect perception as do other forms of theory" (p. 20). Others look at perception from a holistic point of view by suggesting that acceptance or adoption is significantly influenced by innovation characteristics and adopter perceptions (Hahn & Schoch, 1997). Although an innovation is perceived as a new idea, Rogers asserted that sometimes people may have known about the innovation before but may not have developed a favorable or unfavorable attitude toward it, nor have they accepted or rejected it. The newness of an innovation can also be expressed in terms of students who may take an instructional technology course for the first time, may accept or reject its use in the classroom based on knowledge, persuasion, or a decision to adopt it (Rogers, 1995).

As Rogers suggested, the characteristics of an innovation, which is perceived by the members of a social system, determine its rate of adoption. As Rogers further asserted that there are certain innovations spread more rapidly than others. Therefore, the characteristics that determine an innovation's rate of adoption include the following:

Relative advantage is the degree in which an innovation is perceived as better than the idea it supersedes. The key to remember here is when an individual or group perceives an innovation as being advantageous, that innovation may be accepted with a more rapid rate of adoption than an innovation without an advantage.

Compatibility is the degree to which an innovation is perceived as being consistent with the existing values, past experiences, and needs of potential adopters.

Complexity is the degree to which an innovation is perceived as difficult to understand and use. Some innovations are readily understood by most members of a social system; others are more complicated and might be adopted but at a slower rate.

Trialability is the degree to which an innovation may be experimented with on a limited basis. The innovation that has been tested with an assurance of understanding and some sense of confidence will be accepted at a rapid rate of adoption.

Observability is the degree to which the results of an innovation are visible to others. The easier it is for individuals to see the results of an innovation, the more likely they are to adopt it. Such visibility stimulates peer discussion of a new idea (1995, p.162).

Surry and Farquhar (1996) asserted that it summarized a number of variables identified by Rogers, who influenced the rate of adoption. They suggested, "Many other factors, most of them relating to the social factors present at the adopting site, play just as large a role as technological superiority in influencing rate of adoption" (1996, p. 3). Therefore, as shown in Figure 6 below, an innovation's rate of adopting is significantly influenced by the five characteristics of adopter variable based on the perceived attributes of an innovation.

Figure 6. Variables Determining the Rate of Adoption of Innovation

Five Characteristics of Adopters

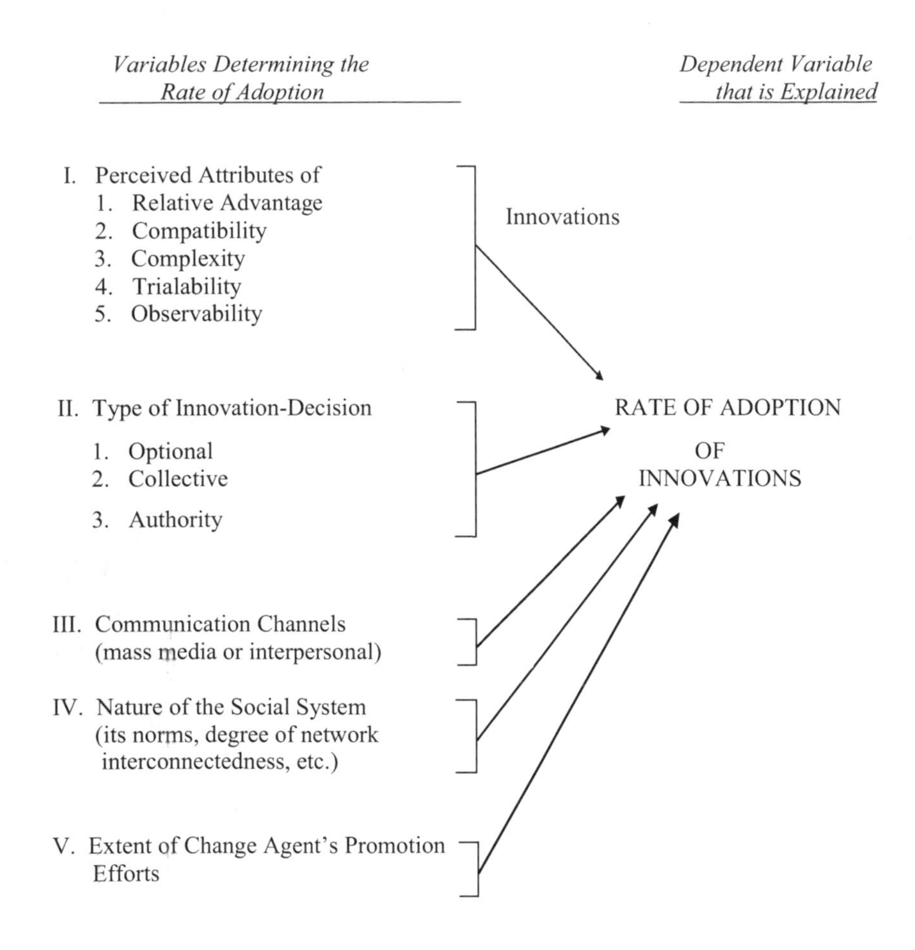

Note: From *Diffusion of Innovations* (p. 207), by E. M. Rogers, 1995, New York: The Free Press. Copyright 1995 by the Division of Simon & Schuster, Inc. Adopted with permission from the author.

Relative to the theory of perceived attributes, Rogers (1995) stated that potential adopters judge an innovation based on their perceptions with regard to the five attributes stated earlier. The theory holds that an innovation will experience an increased rate of diffusion if potential adopters perceive that the innovation can be tried on a limited basis before adoption, offers observable results, has an advantage relative to other innovations, is not overly complex, and is compatible with existing practices and values. The theory of perceived attributes has been used as the theoretical basis for several studies relevant to the field of instructional technology. Other researchers believe that a direct relationship exists between one's attitude about a situation and the individual's perception of the authenticity and relevance of the situation (Simonson & Maushak, 1996, p. 1011).

Another study conducted by Santhanam and Leach (2000) suggested that a gender bias relating to students' perceptions of their ability to use computers showed a significant difference in their responses. Developing confidence in the use of computers seemed to be necessary for some students, particularly female students. The authors' main argument used in favor of computer based media for instruction was to facilitate student learning better than the traditional methods or that they supplement the traditional methods. Their study revealed, "Surveys of student attitudes indicate that students who had prior computer courses or who owned computers were more positive towards computer use in higher teaching/learning than other students (p. 2).

Students' Expectations toward Instructional Technology

The third component is students' expectations toward technology, as affected by the introduction to instructional technology. Students have expectations that higher education will provide them with the tools, techniques, and content needed for productive careers. Students not only bring their prior understanding of instructional technology concepts into the classroom but also they bring a set of attitudes, beliefs, and assumptions about the nature of the class. A study done at five institutions in first semester physics classes showed that many students had expectation misconceptions about the nature of physics and what they should be doing to learn it. The study suggested that the students' expectation rates were to deteriorate rather than improve. The authors concluded, "They come with a need to know what are they to learn, what skills will be required of them, and what they need to do to succeed" (Redish, Steinberg, & Saul, 1996). These "expectations" can affect not only how students interpret class activities but also affect what activities they choose to understand. Among those expectations, students believe they will understand and be able to use information technology more frequently.

According to Hall (1995), students and those already a part of today's workforce need knowledge about technology and skills in its use to remain productive and valued. Among these essential student skills are the basic familiarity and understanding of the role and function of technology. Students need a mastery of technological applications that are germane to their professions and disciplines (p. 28, 1995). They also need a working knowledge of personal computers and multifaceted software tools. In addition, students must possess the ability to search, retrieve, analyze, and use electronic information. In summary, students expect to develop the capability to use technology independently and collaboratively in their chosen fields.

In the information economy, students are demanding the technological skills that they will need to survive in a highly changeable society. Therefore, they not only accept but also demand that essential services be provided to them electronically through (course selection, registration, library catalog, and information retrieval services). Among students, there is a recognized need for lifelong learning and ongoing education. Colleges and universities have the opportunity to define new ways of creating and disseminating information that take into account the changing educational demands of new learners. To meet these demands, Beede and Burnett (1994) argued, "Higher education institutions must evaluate their processes, organizations, and the use of instructional technology and implement a student-centered environment" (p. 69).

In an article of the *Chronicle of Higher Education*, DeLoughry (1996) reported that Green's *1995 Campus Computing* survey revealed that "nearly 24 percent of classes were being held in computer-equipped classrooms and 20 percent of the instructors' courses were using electronic mail" (p. A17). Green's findings indicated that the use of technology for teaching was spreading beyond simple computer use to that of high-tech tools for the enhancement of teaching and learning. The expectations of students have created what Rogers called a "critical mass" since 1995 until now. Green advised that the critical mass of instructional technology was now spreading through higher education. With instructional technology reaching the critical mass, Rogers stated, "What we're seeing this year may even be surpassed by next year" (p. A17).

According to a 1998 survey conducted by Green (1999), most colleges and universities were looking at other ways to reach the critical mass. Today, more administrators are now imposing mandatory charges on student fees to help cover the rising costs of instructional technology (IT). The survey reported that student technology fees had increased about 10 percent in less than a two-year period. Green reported that in 1997, 38.5% of schools reported student fees were used for instructional technology. Contrary to his report in 1995, only 28.3% of student fees were increased. And in 1998, the figure was up to 45.8%. Although some public institutions increased their mandatory IT fees, Green reported, "The average annual fee remained stable at $140 and declined in public four-year colleges from $131 in 1997 to $106 in 1998" (*Campus Computing 1998*, p. 5). According to Green (1998), each year the critical mass of students will be asked to pay additional monies to help alleviate the rising costs of instructional technology. Green's report also revealed that some institutions were beginning to require that entering freshmen come already prepared with a laptop or desktop computer in hand.

In his ninth year of surveying colleges and universities in the United States, Green reaffirmed in his 1998 survey data that computing and information technology were "core components of the campus environment and classroom experience" (p. 3). Green indicated that students came to these universities and colleges expecting to learn about and also to learn with technology. However, his findings indicated that the institutions were struggling to keep up with the demands of students and were unable to strategically plan for the financial burdens they faced, along with planning for curriculum integration and providing for potential user support for students, faculty, and staff. The respondents in this study (mainly the chief academic computing officers, information technology officers, or senior academic officers) revealed several issues of concern for their institutions. When asked to identify the most important issue facing their institution over the next two-three years, 33.3 percent indicated that "assisting faculty [with integrating technology] into instruction" was the number one issue. For the second most important issue, only 26.5 percent reported providing adequate user support (Green, 1998).

Indeed, there has been a significant change in the way institutional officers see their priority needs. Green indicated that the most important concern for institutions was updating their computers, along with purchasing Microsoft computers and Windows 95 operating systems (*Campus Computing*, 1998).

Other researchers reported that the expectations were that the infusion or integration of new technologies into instruction would maintain and enhance student learning while significantly reducing instructional costs (Rakes, 1996; Geoghegan, 1994; and Green, 1996). What most colleges and universities failed to realize was that the majority of students were comfortable with the new technology because accessibility remained a problem for some students. By creating a network of classrooms in real time, only created a teacher-centered form of education, not a student-centered approach (Daniel, 1997; Berner, 1993).

It is obvious that the learning environment is a complex system of mixing both human and technological elements. The foundation of any learning experience is what happens with the learner (student), not what happens with the technology. The driving force in any classroom, whether virtual or concrete, is the human activity and interaction that constitute the learning experiences among students. Some researchers have been able to identify students' learning styles. Carol Twigg (1994) noted an interesting scenario in the Myers-Briggs typology of learners. She concluded that the largest group of college students consist of concrete-active learners who learn best from concrete experiences that engage their senses, that begin with practice and end with theory, while the majority of college faculty prefer the (abstract reflective) pattern, creating an increasing disparity between teacher and learner (p. 24). The use of instructional technology can help to eliminate this disparity. As instructional technology is adopted and courses are modified to accommodate the use of technology in the curriculum, faculty roles will change from content providers to managers and designers of learning environments. Students will also be required to adopt a more proactive and responsible role in their learning experiences. The diffusion of instructional technology is an innovative tool to permeate self-directed learning in the educational social system.

Despite research and testimony that technology is being used by more faculty the diffusion of technological innovations for teaching and learning has not been widespread, nor has it become deeply integrated into the curriculum (Geoghegan, 1994). Geoghegan argued that although there is a growing number of faculty who are very enthusiastic about adopting technology, there is still a large number of mainstream faculty who seem hesitant to adopt technology. He suggests that unrealistic expectations about the development, dissemination, and use of instructional technology can have disastrous consequences for its adoption, especially when these expectations come face to face with the realities of time, money, and skills required for software development, the impact of accelerated technological change, and the difficulty of persuading faculty to incorporate technology into their teaching. (p. 4)

On the other hand, Reeves (1995) questioned whether some researchers, particularly doctoral researchers, would consider social relevance an issue. He argued that one's age, race, gender, socioeconomic status, and education were likely factors to influence one's interpretation of being "socially responsible." He believed that instructional technologists made incorrect assumptions about the nature of instructional technology. He concluded, "Few critics have dealt directly with questions of whether instructional technology research is, can be, or should be socially responsible" (p. 3). In evaluating the success of educational technology, it appears that instructional technology depends largely on how well innovators and early adopters make it work. In order to take advantage of the benefits of instructional technology in higher education, there needs to be an understanding that differentiates early adopters from mainstream adopters. Analyzing student adoption patterns using Rogers's (1995) diffusion theory is a considerable difference from that of students' learning styles.

As Jacobsen (1998) suggests, "Instructional technology may support and increase the efficiency of teaching and learning transactions or even modify educational processes on campus" (p. 3). Therefore, if students' expectations of education are to be met, Jacobsen concluded that instructors will need to become more technologically focused and prepared to deliver education by way of instructional technology. In a recent article of "*The Learning Market Space*," a monthly online newsletter, Bartscherer (2001) noted that there was a need to create learning environments that would allow students to have greater access to technology. She described how students brought multiple interests and expectations to each learning experience. Thus, they differed in the amount of interaction they receive from faculty. As a result, higher education professionals should design more effective online learning environments to accommodate their students' different needs (Bartscherer, 2001).

According to a *2000 Campus Computing Survey*, more college courses were using more technology resources. The report indicated that 59.3 percent of all college courses surveyed, utilized electronic mail, web pages, or other forms of computerized mechanisms (Green, 2001). Another study was conducted at Middle Tennessee State University (MTSU) to examine the effectiveness of teaching using instructional technology. Results of the study helped researchers to determine measures for improving technology resources for the University's faculty and students. Although the survey was administered to only eight percent of the 18,000 student population, results yield very positive responses from an increased student interest and satisfaction rate (Draude & Brace, 1999). Other researchers reported that the expectation of the infusion of new technologies into instruction will continue to enhance student learning while significantly reducing instructional costs (Green & Gilbert, 1995; Rakes, 1996; Geoghegan, 1994; & Green, 1996).

Rogers (1986) also found that the diffusion of technology worked well with individuals in social groups. As he investigated the social impacts of the interactive communication technologies on individuals, organizations, and society, the findings of the effects were the changes in an individuals' behavior (knowledge, attitudes, or actions) that occurred as a result of transmission of a communication message (Rogers, pp. 152-153). In an analysis of social impacts, Rogers used a three-fold typology of influence and consequences interchangeably to mean the changes that occurred in individuals or a social system were a result of the adoption or rejection of an innovation. He called the three components: *desirable impacts* - the functional effects of an innovation on an individual or social system. The second component is called *direct impacts* - the changes in an individual or social system that occur in their immediate response to an innovation; and *anticipated impacts* - changes caused by an innovation that is recognized and intended by the member of a social system. This investigative approach was used to gather data about the social impacts of a technological innovation by comparing a system of certain variables before and after the introduction of a new technology (p. 162).

Wellburn (1996) stated that the use of technology as a learning tool can make a measurable difference in students' achievements, attitudes, and interactions with teachers and other students. She asserts, "Students' attitudes toward learning and students' self-concept were both found to be increased consistently in a technologically-rich environment, and in general, student control was found to be one of the more positive factors relating to achievement when technology was used" (p. 5).

In a recent study conducted by Kuh and Vesper (2001) to learn about the relationship between students' familiarity with the use of computers and other information technologies during college and after, the authors investigated whether becoming familiar with computers contributed to or detracted from the development of skills and competencies such as learning on one's own, thinking analytically and logically, synthesizing ideas and concepts, and working effectively with others. The results of the study showed that students who had increased familiarity in using computers during college significantly contributed to and did not detract from their developmental skills and competencies after college" (p. 95). The study suggested that students who become familiar with computers during college, outscored their low gain counterparts on every outcome measured.

According to Whitehead, Jensen, and Boschee (2003), it is critical that the direction for the technology initiative remain focused on student performance. The authors suggested that the key component behind this program is to create a more symbiotic relationship between technology and the curriculum (p. 175). They believe that by integrating technology into the classroom, it would increase correlations between students' performance and the use of technology to enhance the learning process more effectively. Students participating in interactive technology were more readily independent learners, self-starters, and active collaborators in their experiences spontaneously with greater expectations toward positive learning.

In another study on the key component in the classroom of the future, DeGroot (2002) postulated, "Classrooms have changed dramatically over the last decade with the advent of new technologies and equipment developed to make teaching and learning more diversified and interactive" (p.1 of 4). The author concluded that students no longer had to crowd around a computer monitor to view presentations, web sites, or training programs since multimedia projectors were becoming the centerpiece for classroom technology. In addition, Reed (2003) also believes that it is time to integrate technology throughout the teaching and learning environments. He asserted, "An example of putting the power of technology to work is on-demand video streaming and the promise it has to offer as an instructional tool for a variety of learning experiences" (p. 1 of 4). Furthermore, in a report on the Effect of the "United Streaming Application on Educational Performance" (Boster et al., 2002), students who received instruction incorporating the video-on-demand united streaming application showed dramatic improvements in achievement in three Virginia school districts.

CHAPTER 3

The Theory-Driven Approach

Design Overview

A correlation study was conducted among undergraduate students at two four-year public universities. The study was geared toward a between subjects and theory-driven approach based on Everett Rogers's (1995) theory of the diffusion of innovations. The study was primarily an exploratory investigation designed to gather information from a multidisciplinary group of undergraduate students who were being introduced to a new learning environment through the use of instructional technology. A mixed-method research design was used with a quantitative methodology (selected-response survey items) in conjunction with a qualitative methodology (an open-ended survey response item). In addition, a direct observations was employed for the purpose of investigating the correlation among those students who readily adopted instructional technology for learning purposes in higher education and those who did not. Chapter 3 discussed the methods and procedures used to determine the extent to which students were adapting to changes in their learning environment, associated with the introduction to instructional technology. The study evaluated responses that reflected students' attitudes, perceptions, and expectations toward technology as a means of determining their adoption patterns to instructional technology in the classroom.

To accomplish the proposed study, the investigator adhered to the following procedures:

1. Conducted an exhaustive survey of the literature germane to instructional technology in several areas of higher education.
2. Developed a survey questionnaire which was administered to identified students in the spring semester 2001. The items in the instrument were developed to determine how students were adopting the changes

in their learning environment associated with the introduction to instructional technology. The survey instrument used in the study was given to treathow students' attitudes, perceptions, and expectations toward technology were correlated with an introduction to instructional technology course, with specific adoption categories based on Rogers's (1995) model of the innovation-decision process.

3. Secured permission from administrators at two (2) four-year public Institutions that were included in this study (see Appendix B). The investigator explained the selection of respondents who were asked to participate in the study as shown in Chapter 1. In each case, the necessary assurances of anonymity and confidentiality were given (see Appendix E).
4. Described the instrument to be used in the study, including its validity and reliability, along with the process of the researcher, who collaborated with the committee chairperson to develop the instrument. The researcher provided information on the participants for whom the instrument was prepared.
5. Described the data collection methods to be used in the study, including administrators' permission or consent to conduct the study at the two universities; the researcher obtained informal permission from participating instructors to use students in their classes as participants.
6. Developed a qualitative analysis of the findings as reported in Chapter 4. This chapter served as an overview of the analysis of the actual findings of the study. The researcher analyzed the students' overall behavioral characteristics as a result of the investigation.
7. Developed summaries, conclusions, and recommendations as presented in Chapter 5. This chapter reviewed all the findings and summarized results in order that other researchers may replicate the study. This chapter also served as a model for investigating actual interpretations of students' behavioral patterns and conceptions of instructional technology. Recommendations were presented for future reference in this area of study.
8. Provided a description of the analysis procedures as set forth in the study. Included was a rationale for the statistical methods that were used to identify any limitations or threats to the validity of the study.

Student Population

The investigator conducted the study at Old Dominion University (ODU) and The University of Virginia (UVA),two mid-sized, regional, comprehensive four-year public universities. Old Dominion University has a total population of 18,500 students. The university prides itself on faculty research and distance education with technological innovations such as the TeleTechnet system. This system delivers upper-division undergraduate courses to place-bound students throughout the Commonwealth. Old Dominion University is also the host institution for the National University Teleconference Network (NUTN), a consortium of institutions throughout the country. ODU has been a leader in the development of distance learning technology and applications since 1982. In addition, The University of Virginia has a population of approximately 18,000 students. It has a fully equipped, instructional technology-based systematic application of research and applied learning facility, which prides itself on learning technologies and methodologies for the global economy. UVA addresses the rapidly accelerating changes in the interdisciplinary field of instructional design and assessment, communication arts, business, and educational psychology through the use of instructional technology as a tool in the design and delivery of instruction. Each institution is highly involved in instructional technology pedagogy and has been on the cutting edge in technology for the past fifteen years.

This research surveyed approximately 171 (n=171) of some 300 undergraduate students enrolled in an introductory instructional technology course at Old Dominion (n=150) and The University of Virginia (n=150) for the spring semester 2001. A sample of 171 students were sent letters of consent forms, along with an introduction to the study. Participants received a duplicate consent form as proof of participating in the research project. After reading the letter and consent form, participants were asked to initial and date the form to acknowledge their understanding and agreement with the terms. Participants were informed that the collected data was to be used for dissertation purposes only. They were advised that their identities would remain anonymous and confidential. Only group data was reported (see Appendix B).

After careful research to find a valid instrument to conduct this study, this investigator reviewed research done by Carr, Confessore, and Ponton (1999) based on a Likert scale to measure students' autonomous learning styles with the Inventory of Learner Resourcefulness (ILR), which they developed. Since the instrument used for their study was valid and reliable, the investigator developed a similar Likert scale based on a parametric scale of 0 to 10 points (see Appendix A). However, this parametric scale was used to measure students' behavioral responses to instructional technology.

The instrument used in this study was developed by the investigator entitled, Students' Attitudes, Perceptions, and Expectations Toward Technology (SAPETT) Survey, which is a forty-item questionnaire (see Figure 7). It was used to assess the values of students' attitudes, perceptions, and expectations toward instructional technology based on their self-reported levels of acceptance or rejection in an introductory instructional technology course. The survey instrument was divided into three sections. The first section asked the students for general information such as the name of their college or university, month, day, and year of filling out the survey, and the last four digits of the participant's social security number. Section two asked students to place an x on the scale from 0 to 10 to indicate how often the item applied to their experiences in an introductory instructional technology course. A score of 0 on the ranking scale meant that the statement was never descriptive of them. A score of 5 meant that the statement was sometimes descriptive of them. A score of 10 meant that the statement was always descriptive of them. Participants were asked also to identify any past experiences they had with any form of instructional technology such as video conferencing, online group discussions, chat sessions, or electronic mail systems. Section 3 asked participants for demographic information, including the students' gender, classification, socioeconomic status, age, and number of credit hours taken in instructional technology courses. An open-ended question was placed last for students to describe their ethnicity (see Survey Instrument).

Survey Instrument

Students' Attitudes, Perceptions, and Expectations Toward Technology (SAPETT) Survey

1. I feel that I the instructional technology course will provide me with an interactive learning experience with other classes.
2. I like group discussions in a new learning environment.
3. I believe that instructional technology will help to enhance my learning experiences in class.
4. I feel that the electronic classroom environment will provide a positive learning experience for me.
5. I feel instructional technology is a convenient means of acquiring new knowledge.
6. I think this classroom environment will help improve my grades.
7. I like this course because there are no set restrictions on my response time to class assignments.
8. I think this class will be a good learning experience for me.
9. I enjoy coming to this class because I like learning new things.
10. I feel that the new learning environment in the instructional technology class will not restrict my learning abilities.
11. I feel that I am learning more from this class.
12. For the courses I am enrolled in this Spring 2001 semester, I send e-mail communication to fellow students.
13. For the courses I am enrolled in this Spring 2001 semester, I send e-mail communication to instructors/professors.
14. I feel that writing e-mail messages will improve my writing skills.
15. I am able to work effectively in groups when completing course assignments.
16. I am able to initiate effective working relationships with fellow students when studying for exams.
17. I am able to have access to a computer off campus.
18. I am able to connect to the university's network from off campus.
19. I am required to learn in a seminar setting environment.

20. I think this introductory class of instructional technology should be a required course.
21. I feel that the new learning environment in the instructional technology class will not restrict my learning abilities.
22. I want an integrated learning environment with access to other knowledge outside the classroom environment.
23. Instructional technology is a tool for me to have group discussions with other students in other classes.
24. I see technology as an unlimited source for gathering information for my use as well as sharing information with others anywhere and at any time in the world.
25. I believe instructional technology is an expensive way to continue my education.
26. Taking classes in instructional technology will enable me to become a better student.
27. The introduction to instructional technology course is helping me to make use of my time.
28. For the courses I am enrolled in this Spring 2001 semester, I have access to course materials through the Internet.
29. I am required to learn in a large lecture environment.
30. With the use of instructional technology, I expect the instructor to use other technology tools to broaden my knowledge.
31. With the use of instructional technology, I expect to be able to communicate with my instructors.
32. I am satisfied with the instructions I have received from the instructor/ professor regarding the process of the course.
33. I am satisfied with the instructor's method of teaching.
34. I am satisfied with the instructor's fairness in grading my work.
35. I am satisfied with the overall instructional environment.
36. I am satisfied with the expectations of the instructor.
37. I am satisfied with the expectations I have of the introduction to instructional technology class.
38. I expect that this course will enable me to be prepared for future job opportunities.
39. The class is everything I expected and more.
40. I would recommend this course to other students.

DEMOGRAPHICS:

Gender: Male ____ Female ____

Classification: Freshman ___ Sophomore ___ Junior ___ Senior ___

Annual family income:
(less than $25,000) ___ ($25,000-$69,999) ___ ($70,000 or more) ___

Age: ____ Exposure to Instructional Technology: ______________
(college credit)

Describe yourself in terms of ethnicity in one or two words:

____

____

Students had an opportunity to express some general feelings, beliefs, and general impressions regarding their behavioral characteristics toward instructional technology associated with their new learning environment. Some questions asked students to rate their perceptions about various types of instructional technology methods used in their teaching and learning environments. Other questions were used to measure students' expectations about instructional technology and the impact it has on their learning. The instrument used for this study was based on a Likert scale similar to the Inventory of Learner Resourcefulness (ILR) developed by Carr, Confessore, and Ponton (1999) proved to be valid and reliable for this particular study. This investigator developed the same Likert scale of 0 to 10 to measure students' responses to instructional technology. Each undergraduate participant was asked to rate the statements based on literature research for this study.

Johnson (1989) found that the use of discussion questions followed by a mediated situation resulted in significant attitude changes toward careers with regard to learners' confidence in their ability to be successful. Also, Ehrmann (1997) suggested that instructors who use instructional technology as a method of teaching, "focus on the knowledge skills and wisdom that students need in order to perform well in their new learning environment after graduating from college (*Change*, p. 23).

Kosakowski (1998) revealed that the application of educational technologies in instruction has progressed beyond the traditional rote memory to include the use of complex multimedia advanced networking technologies. He found that the innovative approach of students using technology, allowed students to have more control over their own learning environment. Kosakowski argued, "Traditional lecture methods are often left behind as students collaborate and teachers facilitate" (*ERIC Digest*, ED 420302).

According to Rogers (1995), an individual's decision to adopt an innovation is not an instant act. Rather, it is a process that occurs over time based on actions and decisions that are made by the individual. Rogers' model of the innovation-decision process is "the process through which an individual passes from first knowledge of an innovation to forming an attitude toward the innovation, to a decision to adopt or reject, to implementation of a new idea, and to a confirmation of this decision (p. 163).

As Fitzelle and Trochim (1996) suggested in their study at Cornell University, a sample of undergraduate students enrolled in a research methods class thought that the web site called Scavenger Hunt significantly enhanced their learning of course content. The authors concluded, "students' perceptions of performance in the course were predicted by variables of enjoyment and control of their learning pace" (p. 1). In a recent study conducted by Bialo and Sivin-Kachala (1996), other benefits enjoyed by students who used technology revealed positive attitudes toward self and learning. This study suggested that students felt more successful in school, were more motivated to learn, and had increased self-confidence and self-esteem after having experiences with computer assisted instruction (CAI). A similar study found that CAI significantly improved student achievement and attitudes toward technology while decreasing instruction time (Ester, 1995).

To meet the new demands of students, Beede and Burnett (1994) argued, "Higher education institutions must evaluate their processes, organizations, and the use of instructional technology and implement a student-centered environment" (p. 69). Johnson and Liu (1998) found that students' attitudes toward information technology were a major factor affecting their success in learning and using technology. The authors conducted a study which compared the attitudes of elementary education major students with that of secondary education major students' attitudes toward information technology. Their study revealed four major variables: (1) enjoyment - the degree to which students enjoyed learning and working with a computer; (2) motivation - the degree of motivation students have when they learn and use the computer; (3) importance - the extent to which students see learning and using the computer, and (4) freedom from anxiety - the degree to which students feel anxious while learning and using the computer. The results of the study revealed that the secondary education students had a significant difference in attitudes toward information technology than the elementary education students.

According to Simonson and Maushak (1996), "A direct relationship exists between an attitude about a situation and the individual's perception of the authenticity and relevance of the situation" (p. 1011). Other research indicates that individuals in organizations develop perceptions of what they are entitled to receive in exchange for their contribution. They must perceive that they are being treated equally as others in their group. The expectations of students have created what Rogers (1995) calls the "critical mass" which is spreading through higher education (DeLoughry, 1996). As a result of this critical mass of students, Kenneth Green (1998) reported that student technology fees had increased to 45.8 percent, a profound increase since 1980 (*Campus Computing*, 1998).

Rogers asserted that sometimes people may have known about the innovation before but may not have developed a favorable or unfavorable attitude toward it. He concluded that the newness of the innovation can also be expressed by students who have enrolled in an instructional technology class for the first time and may not accept or reject its use based on knowledge, persuasion, or a decision to adopt it (Rogers, 1995). His theory of perceived attributes, potential adopters judge an innovation based on their perceptions with regard to five attributes: trialability, observability, relative advantage, complexity, and compatibility. The theory maintains that an innovation will experience an increased rate of diffusion if potential adopters perceive that the innovation can be 1) tried on a limited basis before adoption; 2) observable with results; 3) an advantage relative to other innovations; 4) easy to understand and not overly complex; and 5) compatible with existing practices and values.

Surry and Gustafson (1994) conducted a study to determine the role that perceptions played in the adoption of an instructional innovation. They argue that an innovation is more likely to be adopted if potential adopters have favorable perceptions with regard to its complexity, compatibility, relative advantage, observability, and trialability. Their research design involved the use of questionnaires and interviews to determine the perceptions of the potential adopters who used computer-based learning. Results showed that those who favored the adoption consistently used compatibility, complexity, and relative advantage as their choice of Rogers's perceived attributes. The researchers concluded that these attributes could be important considerations when introducing an innovation into instructional settings (Surry & Gustafson, 1994).

One researcher concluded that the cost of taking a course in instructional technology can become a barrier when one considers accepting new technology. However, he cautions, "What is frequently not understood is the cost of not adopting new approaches." Garland notes a prime example that justifies the perception that some people have toward technology: If one hour of paid student time can be saved in a course and that hour, multiplied over hundreds or thousands of students, exceeds the cost of the technology needed to save the hour, then the use of the technology may be justified. (1995, p. 284)

In a recent article of *Syllabus*, a monthly magazine about using technology as a teaching and learning tool, Mesher (1999) presents the reasoning and schema scripting interactivities into courseware as a challenge to improve teaching and learning in the classroom. He asserts, "Universal interaction with students, combined with accessibility at the student's convenience, exact repeatability, and uniform quality, gives asynchronous online learning the potential, in suitable situations, of not merely replacing, but surpassing the classroom learning experience" (p. 20). Mesher supported interpersonal interactivity by suggesting that various behavioral characteristics play a major role in students' expectations when asynchronous projects on the Internet such as e-mail, listservs, chatrooms, or bulletin boards can help to reproduce collaborative peer learning in the classroom. Through the use of these pedagogical devices, online communication will give students more direct access to the instructor as well as other students at any time, and anywhere in the world (p. 23). Therefore, using technology to complement instruction extends beyond the instruction of the teacher and it provide learners the opportunity to practice skills in private, and it promotes self-direction by allowing learners to supplement instruction in ways that meet their individual needs. However, Rogers asserted that individuals tend to expose themselves to ideas that are in accordance with their interests, needs, and existing attitudes. They consciously or unconsciously avoid messages that are in conflict with their attitudes and beliefs (Rogers, 1995, p. 168).

According to William Geoghegan (1998), faculty members have not been able to reach a saturation point because they have not committed themselves to instructional technology. He suggested that only five to 15 percent of faculty members nationwide, have reached the "so-called early adopters" category. Geoghegan credits his findings to research done by Everett M. Rogers, a social science theorist in the field of social science and communication. Geoghegan asserted that Rogers's theory of the diffusion of innovations, "seeks to describe and understand the processes that underline the acceptance and spread of innovations among members of a community" (p. 134). Geoghegan believes there is a pattern that exists whereby individuals adopt innovations over time through a "normal bell-shaped distribution, or the adoption process moves through a predictable sequence of well-defined adopter categories or groups based on social and psychological characteristics" (p. 134).

Rogers (1995) referred to the groups as innovators- 2.5 percent, who are the first to adopt an innovation; early adopters - 13.5 percent has the greatest degree of opinion leadership; early majority - 34 percent and late majority 34 percent (these are the "critical mass"); and laggards - 16 percent, the last to adopt (p. 262). A strong advocate for theory development in the field of instructional technology, Charles Reigeluth (1996) is a practitioner who has experienced the historical evolution of research in the field of instructional design strategies (IDS). In one of his recent articles, "A New Paradigm of IDS?" documents his interest in theory development and its role in restructuring education. Reigeluth believes, "the learning-focused instructional theory must offer students guidelines for the design of learning environments that provide appropriate combinations of challenge and guidance, empowerment and support, self-direction and structure" (p. 15). He argues that since the world has become more complex, students need more skills for complex cognitive tasks such as solving problems in non-structured domains. The need for instructional-design strategies is basically a process for making decisions about the nature of instruction and of ways to facilitate learning through analysis, synthesis, and evaluation. Reigeluth stresses these three issues in instructional technology because "performance problems almost always require organizational changes as well as changes in the knowledge and skills of individuals" (*Educational Technology*, p. 15). To help students achieve their potential goals, Reigeluth advises educators to customize students' learning capabilities instead of standardizing their learning process. He concludes that higher education constituents curtail the thought of deleting the traditional method of teaching; instead, they should "incorporate most of the knowledge the field has generated about both instructional theory and the IDS process" (p. 19).

Garland (1995) asserted that in colleges and universities, "technology can be used to individualize lessons, thus improving the quality of learning and for some students, increasing the amount of education available in a given time frame" (p. 283). Thus, the use of instructional technology associated with the introduction to instructional technology is essential when considering students' styles of learning in a new learning environment. According to Simonson (1995), educators have long been concerned about students' attitudes toward learning. He suggested that although students' attitudes have not been scientifically linked to one's achievement, "It should be considered as a vital element of effective instruction" (p. 365. Also, as the demand for higher education to become more competitive in a global society, Saba (1998) suggested that faculty members, "become actively involved in shaping the economy of the future so that students can become more independent in their learning environment" (*Distance Education Report*, pp. 2-5). Instructors of the classes did not use this survey as part of the students' final grades for the course. It was the intent of the investigator to use the survey instrument findings strictly for dissertation purposes.

Research Procedures

At the beginning and the end of the spring 2001 semester, students enrolled in an introductory instructional technology course were administered a questionnaire. They were asked to evaluate their classroom experiences with the integration of instructional technology. In order to conduct this study, the researcher considered several factors. The investigator conducted the following: (1) Sent a letter of request to obtain permission from the vice president of academic affairs or designated office to conduct this study. Acquired cooperation from each of the selected institutions to ensure that the investigator's ethical standards adhere to the institution's policies and procedures. The investigator's letter of request explained the study's purpose and ensured that the participants' rights of confidentiality were respected throughout the study. (2) Sent a letter of request to three professors at the selected institutions to request their participation in the study and their permission to conduct the study. The investigator included in the letter all the necessary preparations for the study, addressed the issue of ethical standards, protection of students' rights of confidentiality, and submitted a copy of the survey questionnaire and student consent form. The investigator kept in close contact with the professors for the semester. The investigator mailed and/or delivered the materials for the survey to each university professor. The survey was conducted at the beginning of the spring 2001 semester. Upon the participants' completion of the survey, each professor was asked to return the survey information to the investigator in a self-addressed, stamped envelope. The investigator also included copies of letters to both professors and students. (3) Sent e-mail and written correspondence to professors at the three universities, explaining the nature of the study. Included in the letter was a statement of purpose of the study, along with procedures for distributing and collecting the questionnaires. (4) Sent out survey correspondence to participating instructors at each university to disseminate to 40 or 50 students in each class (total 250-300 student participants). Included a self-addressed, stamped envelop for return mail to the investigator, as well as an invitation to send an abstract of the study upon completion of the project. (5) Mailed correspondence, questionnaires, and consent forms through the U.S. Post Office, first-class mail, or hand-delivered packages to designated professors to ensure proper delivery. (6) Placed number codes on the top right corner of the receipt control of questionnaires and initialed consent forms. (7) Limitations or threats to the validity stemming from the study were based only on information obtained from the two universities in southeastern Virginia.

Personal Observation

The investigator of this study visited the two public universities and conducted a personal observation of the actual classrooms at (ODU) Old Dominion University in July 2000 and March 2001, and at The University of Virginia (UVA) in March 2001. Two classes were observed at ODU: one class during the weekday and one class on a Saturday morning. Well-organized and very informative professors taught all three classes. The three professors used "instructional technology" as a method of teaching with the use of multimedia technology (teleconferencing with other classes outside of the university, PowerPoint presentations, CD-ROM with voice activation, etc. During the observation, the investigator noticed that the students were either busy taking notes, observing the professor on a monitor(s) or looking at other equipment in the room while listening to other students from other universities or surrounding sites that were not visible for the students inside the main classroom to see. The professors lectured extensively and at times would call attention to a student inside the classroom or talked to another student through a monitor who could not be seen by the outside students. The professor was not able to see other students outside the main classroom, but he could use a censoring device to answer questions from students outside the classroom. The students outside the classroom were able to see the professor and the students in the classroom with the professor. However, the students in general did not communicate with each other.

Within the main classroom setting, the students did not seem to communicate with each other as well. Some students in the Saturday morning class barely knew each other. While waiting for the professor to come to the Saturday morning class, the investigator asked the students if they were always as quiet as they were. They responded slowly with one individual who seemed relieved that someone asked him a question. That question prompted others to join in the conversation with smiles and chuckles. One student had communicated to another student on a group project through e-mail, but the two students did not know what the other one looked like until the question was asked. The investigator found this observation to be surprising that even though some students were working together on a project assigned by the professor, they did not know each other officially or had communicated with each other during class time. Other students in both classes seemed distant from each other and there was little time for conversations within the classroom.

At UVA, one professor used a PowerPoint presentation while giving a lecture to a class of some 75 students. The subject matter was about how other countries celebrated the Easter holiday, which was very interesting and the PowerPoint slides were enlightening. However, in observing most students, they seemed preoccupied by small conversations with each other. Occasionally, the professor made a striking comment that gained the students' interest with a laugh from humorous comments made by the professor or another student. While making these observations from the three classrooms, the investigator noted that although students liked to be entertained with visual aids, it was quite obvious that they also lacked the one-on-one conversations with their professors during the class. Thus, personal observations led the investigator to believe that there is no significant difference between students' attitudes and perceptions toward instructional technology. However, there seemed to be a significant difference in the students' expectations toward instructional technology. Students still want personal communication with their professors as well as with other students either inside or outside the classroom.

Data Collections

Surveys were mailed to selected professors at two, four-year public universities for distribution among their students who were currently enrolled in introductory instructional technology courses. In order to ensure that at least 50 percent of the surveys were returned for evaluation, this investigator sent a letter to the vice president or designated official explaining the need for conducting this study and requested permission to conduct it (see Appendix A). Each administrator was sent a copy of the participant's letter of request and a consent form, as well as the survey for information purposes and approval. Upon official approval from each administrative officer, the investigator contacted four professors who were currently teaching a course in the use of instructional technology. Each professor received a letter of request (see Appendix B) and a copy of the student's letter of request, consent form, and the survey. Each professor agreed to participate in this study.

Once the investigator obtained approval and agreement to participate from each professor, final copies of the first surveys were mailed to each professor (approximately 150 copies for each university). The investigator sought the help of at least three professors at each school to ensure a 50 percent return rate. The survey was sent out at the beginning of the spring semester (see Figure 7). Participants were asked to rank their perceived levels of experience in instructional technology and indicate their reactions to the diffusion of innovations with the use of electronic mail communication, video conferencing, and overall exposure to the increasing array of instructional technology. Completed surveys were returned by 171 respondents from the two universities; seventy-seven responded from University A and ninety-four responded from University B. The sample was diverse in terms of age, gender, classification, and ethnicity. Ethnicity was addressed in an open-ended statement and was summarized in Chapter 4.

Validity of the Instrument

The SAPETT Survey was developed by the investigator based on research done by Carr (1999) and Ponton (1999). The two researchers constructed an instrument called the Inventory of Learner Resourcefulness (ILR), a 10-

point Likert scale to accurately assess respondents' intentions to undertake the behaviors associated with learner resourcefulness in autonomous learning. Also, a group of professionals and colleagues in the area of instructional technology were asked to examine the survey and give constructive feedback. Each reviewer was asked to read each item of the survey and give a candid assessment of whether or not the items on the survey would generate the kind of information needed for this study. Revisions to the SAPETT Survey were made based on suggestions and comments from the reviewers and observations from the field test. In addition, a pilot test was done with a group of 15 students enrolled in a summer course of instructional technology who had prior experience in this field at the selected institutions, and they were asked to complete the survey as a pilot-testing procedure. The pilot test of students' responses on the survey showed that students did, in fact, understand the survey statements.

CHAPTER 4

Students' Ability to Adapt to Change

Analysis of Data

Data from the survey were treated on the computer using the Statistical Package of Social Sciences (SPSS) statistical software package. Data relative to students' attitudes and acceptance were subjected to the Pearson Correlation Statistical Sub-test. The level of significance was set at 0.05. Results tended to support the research hypotheses. This study was conducted to show several variables related to college students' attitudes, perceptions, and expectations toward the use of instructional technology. It also emphasized their ability to adapt to changes in their learning environment associated with the introduction of instructional technology. The dependent variables to be considered in this study included gender, classification status, age, years of experience in a computer technology environment, attitudes, perceptions, and expectations. The main independent variable in the study was innovativeness.

Innovativeness is a continuous variable because it captures discrete categories as a conceptual device. It is advisable that when investigators are considering the standardized adopter categories, they must decide on three important factors: 1) The number of adopter categories, (2) The portion of the members of a system to include in each category, and (3) The method of defining adopter categories. In order for these steps to be valid, a criterion for adopter categorization has to be in place. As a result, categorizing innovativeness merely simplifies the classification process, which aids in the understanding of human behavior. To note the distinction of classifying overt behavioral patterns, a set of categories will include various units of the study; each category is mutually exclusive or excludes a unit of study that appears in one category form, and each category must be derived from a single classification principle.

Adoption Distribution of Respondents

When considering the statistical factors, the mean (x) and the standard deviation (*sd*) are the two factors used to divide a normal adoption distribution into its various categories:

Category	Classification
0-80 = Laggards	5
81-160 = Late Majority	4
161-240 = Early Majority	3
241-320 = Early Adopters	2
321-400 = Innovators	1

The ages of the respondents ranged from 18 to 44. Both genders were included for freshmen, sophomores, juniors, and seniors who completed and returned questionnaires. The data indicated that there were no freshmen who took part in the survey at Old Dominion University. There were twelve respondents in the 22+ group from Old Dominion University and thirteen respondents in the freshman category were from The University of Virginia. However, the data showed that most of the respondents (51) who were juniors between the ages of nineteen and twenty-two-years-old were from University of Virginia (see Tables 1 and 2).

Table 1- Number of Participants who completed the survey

Description of Sample Old Dominion University						
Age	**18**	**19**	**20**	**21**	**22**	**22+**
M/F	**M F**	**M F**	**M F**	**M F**	**M F**	**M F**
Class.						
Freshmen	**0 / 0**	**0 / 0**	**0 / 0**	**0 / 0**	**0 / 0**	**0 / 0**
Sophomores	**0 / 0**	**1 / 0**	**0 / 1**	**1 / 0**	**1 / 0**	**0 / 1**
Juniors	**0 / 0**	**2 / 2**	**7 /11**	**3 / 8**	**2 / 1**	**1 / 4**
Seniors	**0 / 0**	**0 / 0**	**3 / 2**	**2 / 9**	**3 / 9**	**3 / 3**

Table 2- Number of Participants who completed the survey

Description of Sample University of Virginia						
Age	**18**	**19**	**20**	**21**	**22**	**22+**
M/F	**M F**	**M F**	**M F**	**M F**	**M F**	**M F**
Class.						
Freshmen	**10/ 3**	**5 / 0**	**0 / 0**	**0 / 0**	**0 / 0**	**0 / 0**
Sophomores	**0 / 0**	**0 / 3**	**0 / 0**	**0 / 0**	**0 / 0**	**0 / 0**
Juniors	**0 / 0**	**2 / 7**	**18/16**	**5 / 2**	**1 / 0**	**0 / 0**
Seniors	**0 / 0**	**0 / 0**	**0 / 2**	**4 / 8**	**2 / 3**	**1 / 2**

Students' Attitudes and Acceptance Data

Subhypothesis #1 showed a positive correlation between students' attitudes toward the diffusion of technology into the instructional program and their acceptance of instructional technology. In its null form, subhypothesis #1 states: There is no correlation between students' attitudes toward the infusion of technology into the instructional program and their acceptance of instructional technology. The data relative to attitude and acceptance were subjected to the Pearson Correlation Statistical Subtest. The level of significance was set at 0.05. Results revealed an extremely high correlation between attitude and acceptance, which was significant at 0.01; the correlation was .946. Consequently, there was a positive correlation between students' attitudes and their acceptance of diffusion. A summary of these analyses is illustrated in Table 3.

Table 3- Summary of Analysis of Attitudes and Acceptance

Correlations

	Attitudes	TOTAL
Attitudes		
Pearson Correlation	1.000	.946**
Sig. (2-tailed)		.000
N	171	171
TOTAL		
Pearson Correlation	.946**	1.000
Sig. (2-tailed)	.000	
N	171	171

Note. Correlation is significant at the 0.01 level (2-tailed).

Students' Perceptions and Acceptance Data

Analysis of Data Relative to Subhypothesis #2 as stated in Chapter 1 revealed: There was a positive correlation among students' perceptions toward the infusion of technology into the instructional program and their acceptance of instructional technology. In its null form, subhypothesis #2 states: There was no correlation among students' perceptions toward the diffusion of technology into the instructional program and their acceptance of instructional technology.

The acceptance and perceptions data were subjected to the Pearson Correlation Statistical Subtest. The results in this case were similar to those in hypothesis #1. A correlation of .899 with a probability <.01 was revealed in this study. Therefore, the null hypothesis was rejected and the research hypothesis is accepted. Apparently, there was a positive correlation between students' perceptions and their acceptance of diffusion as indicated in Table 4.

Table 4- Summary of Analysis of Perceptions and Acceptance Data

Correlations

	Perceptions	TOTAL
Perceptions		
Pearson Correlation	1.000	.899**
Sig. (2-tailed)		.000
N	171	171
TOTAL		
Pearson Correlation	.899**	1.000
Sig. (2-tailed)	.000	
N	171	171

Note. Correlation is significant at the 0.01 level (2-tailed).

Students' Expectations and Acceptance Data

There was a positive correlation among students' expectations toward the diffusion of technology into the instructional program and their acceptance of instructional technology. In its null form, subhypothesis #3 states: There was no correlation among students' expectations toward the diffusion of technology into the instructional program and their acceptance of instructional technology. Analysis of these data, using the Pearson Correlation Statistical Subtest resulted in the following data: The correlation was .899, with a probability of <.01, and there were 171 participants. As with the other two variables, there was a positive correlation between expectations and acceptance. The null hypothesis was rejected and the research hypothesis was accepted. Table 5 summarizes these findings.

Table 5 -Summary of Analysis of Expectations and Acceptance Data

Correlations

	Expectations	TOTAL
Expectations		
Pearson Correlation	1.000	.989**
Sig. (2-tailed)		.000
N	171	171
TOTAL		
Pearson Correlation	.898**	1.000
Sig. (2-tailed)	.000	
N	171	171

Note. Correlation is significant at the (0.01) level (2-tailed).

Students' acceptance of moving technology into the instructional program was influenced by certain identifiable factors (variables). In its null form, the major hypothesis states: Students' acceptance of moving technology into the instructional program was not influenced by certain identifiable factors (variables). Based on the fact that each of the three subhypotheses were found to be true, it was concluded that the major hypothesis was also true. The null hypothesis was rejected and the major hypothesis was accepted. This investigator found that there was a positive influence between acceptance of the diffusion of technology into instruction and the identified variables.

Summary of Respondents

In summary, approximately 300 surveys were distributed to two Virginia public universities, hereinafter referred to as Old Dominion University (ODU) and the University of Virginia (UVA). Seventy-seven of 150 surveys were completed and returned from Old Dominion University for a return percentage of 48. Ninety-four of 150 surveys were completed and returned from the University of Virginia. The results from the University of Virginia revealed a return percentage of 66. Overall, out of a total of (171/300), there was a 57 percent return rate. Summarizes of the data with in-depth explanations on the investigator's findings are inclusive in Table 6.

Table 6 - Summary of Respondents by Acceptance Category

Number of Students and Percentages

	IN	EA	EM	LM	LA
ODU	8/10.39	41/53.25	22/28.57	6/7.79	0/0
UVA	15/15.96	52/55.32	25/26.59	2/2.13	0/0
Total	23/13.45	93/54.39	47/27.49	8/4.68	0/0

Note. The results show there are no laggards at either university in instructional technology.

After observing the students in class at the two universities and analyzing the data, this investigator found with no surprise that any of the respondents were identified as laggards (LA), and most of them (93/54.39%) were in the early adopter (EA) category. Forty-seven of them (27.49%) were early majority (EM). Twenty-three respondents (13.45%) were categorized as innovators (IN). The other eight (4.687%) fell in the late majority (LM) category. The numbers and percentages at the universities were closely related in overall figures.

At Old Dominion University, the data indicated that eight respondents (10.39%) were innovators; there were forty-one (53.25%) early adopters. Twenty-two of them (28.51%) were early majority, and the other six were (7.79%) were late majority. As a result of the other findings, there were no (0%) laggards among the group. Also at the University of Virginia, the research indicated that fifteen of them (15.96%) were in the innovator category, and fifty-two respondents (55.32%) were in the early adopter category. Twenty-five of them (26.59%) were early majority. The other two (2.13%) were categorized as late majority. Again, there were (0%) no laggards among this group.

The SAPETT questionnaire had one open-ended item which asked respondents to describe themselves in terms of ethnicity in one or two words. Some interesting responses were generated by this item. As a matter of fact, some of the respondents had fun with this item. Their responses were intriguing as well as humorous. For example, the Caucasian and White responses and the African-American and Black responses were fairly even. The somewhat odd responses resulted in some chuckles. The investigator compiled a list of responses given by the students at the two universities. The numbers in parenthesis are the frequencies at which the responses occurred as follows in the area of ethnicity as shown in Table 7 and Table 8 below.

Table 7

ODU - Ethnicity Responses	
Caucasian (15)	Mexican-American
White (13)	Russian/German
African-American (9)	African
No Response (9)	Irish-American
Black (3)	Caucasian-American-Indian
Hispanic (3)	Filipino-American
Black Hispanic	European-American
White Hispanic	Native American
White Anglo	Mixed
Diverse	Original Cracker (White)
Anglo-Saxon	Tall
American Citizen	WASP
White Boy	Badass
Beer Drinker	

Note: Responses given based on students' answers to the open-ended question about ethnicity.

Table 8

UVA - Ethnicity Responses	
White (31)	Chinese-American
Caucasian (23)	Non US
Asian (6)	Snow Flake
No Response (4)	Indian
Asian-American (3)	Indian from India
African-American (3)	Korean
Black (2)	Asian Indian
Anglo-Saxon (2)	Human
1/2 Chinese 1/2	American-Persian
White (2)	Other
Homo Sapien	Indian-American
Hapu-Itanli	Myself
Filipino	Hispanic
Jewish	Mixed
Messed up	Middle Easterner
White Caucasian	
White Skinned	

Note: Responses given based on students' answers to the open-ended question about ethnicity.

CHAPTER 5

Conclusive Evidence

Summary of Findings

Subsequently, technology is an important part of many aspects of life and education is no exception. Of course, if education is not to be left behind, we must somehow use technology in our instructional programs. If this is to be successful, we must make certain that our students are amenable to the use of technology in instruction. The current study was pursued with these thoughts in mind. A review of related literature established that many educational institutions are beginning to use technology in instruction as a mainstream innovation; therefore, it behooves us to make a concerted effort towards the inculcation of technology with instruction.

Our examination of Rogers's (1995) theory of the *Diffusion of Innovations* as it was manifested at two state universities in Virginia, suggested that there was a positive relationship among students' attitudes, perceptions, and expectations toward instructional technology in relation to the diffusion of innovations. This study indicated a very high correlation between acceptance of diffusion and students' attitudes, acceptance of diffusion and students' perceptions, and acceptance of diffusion and students' expectations.

After examining the subjects of this study in terms of Rogers's theory, the study found no respondents who could be categorized as laggards. As a matter of fact, most of them were early adopters (54.39%). The next largest category was early majority with (27.49%), and a fairly large percentage were innovators (13%). A good number were easily majority, and a few were late majority (4.68%). Given these observations, it appears that instructional technology is a concept whose time has arrived. In addition to that, the high correlation which was generated by the Pearson Correlational Statistical Subtest suggests that the student component of the higher education community was readily accepted for such a development.

Implications

Students' attitudes, expectations, perceptions, and their acceptance of the concept of diffusion also suggested that instructional technology should be incorporated into the instructional programs of all colleges and universities. The veracity of the major hypothesis was borne out by the fact that each of the subhypotheses were accepted. A high correlation among students' attitudes, perceptions, expectations, and acceptance suggest that the diffusion of instructional technology should be accepted with considerable ease.

Acceptance of subhypothesis #1 indicated that students' attitudes toward diffusion were highly positive. Therefore, students will accept diffusion readily. Acceptance of subhypothesis #2 suggested that students' perceptions of the diffusion of instructional technology into instructional programs of universities are favorable to that development. They expected more insight in the learning process with the diffusion of instructional technology into instructional programs. Another indication of the amenability of the respondents to diffusion was provided by the acceptance of subhypothesis #3. Outcomes of this study gave clear indication to substantiate Rogers's (1995) Theory of the Diffusion of Innovations.

One scenario of this study concluded that there was a high correlation among students' attitudes and acceptance of the diffusion of technology into instructional technology programs. The study also concluded that there was positive correlation among perceptions and acceptance of diffusion of technology into instructional technology programs. This correlation was also extremely high at (.899). This study found that there was an extremely high correlation among students' expectations and acceptance of the diffusion of technology into instructional technology programs. Considering the high correlation between the dependent variables and acceptance, the diffusion of instructional technology as viewed here should not threaten the teaching and learning process in higher education. In the event that difficulty is encountered when attempting to initiate an instructional technology program, this study indicated that attention to students' attitudes, perceptions, and expectations of technology should facilitate such initiation. If those who are instrumental in diffusing technology into instruction would take measures to improve students' attitudes toward technology, to improve students' perceptions of technology, and to improve students' expectations of technology, the innovative process would reach full saturation among colleges and universities.

Recommendations

Based on the findings of this study, I would recommended that the educational institutions make some effort to discover the levels of their students' attitudes, perceptions, and expectations as related to the diffusion of technology into instruction. Furthermore, I think that educators should implement various programs designed to improve students' attitudes, perceptions, and expectations toward technology. If this action in implemented as a tool of instructional technology into colleges and universities on a more global level, the diffusion of technology into instruction will be greatly facilitated. During the course of this study, other questions were raised about other variables and their relationships with acceptance. One such variable was gender. I would also recommend that other research should be conducted to investigate those relationships. The first recommendation is that more instructional technology should be included in the instructional programs of institutions. I would suggest that such an effort would encounter only minimal resistance from students since the majority of them are already exposed to or are highly skilled in computer technology. If we accept Rogers's theory of the diffusion of innovation with instructional technology, any difficulties which might arise would be eliminated by addressing the attitudes, perceptions, and expectations of students. This research showed that students were highly amenable to instructional technology. Students are more accepting of instructional technology and would welcome the change and challenges they might encounter to learn from a broader perspective rather than from a narrow focused concept.

A good deal of trepidation has been observed by this investigation among faculty, students, and staff at colleges and universities. Therefore, I would recommend that an effort similar to the current study as it relates to faculty and staff be conducted as well. In addition, one might also be curious to know how the diffusion of technology into instructional programs would be embraced across ethnic lines as well as other fields of study. The socioeconomic status (SES) could have an impact on whether or not individuals would be receptive of technological instruction. The socioeconomic status most certainly would have a profound influence not only on an individual's access to technology but also on the way in which instructors use technology as a method of delivering education as innovative change agents.

Glossary of Terms

Anchored instruction: a method of using technology to provide a realistic situation for learning.

Change agent: an individual or a team who succeeds in understanding the organization, its culture, its resources, and its politics.

Communication: a process in which participants create and share information with one another in order to reach a mutual understanding.

Compatibility: the degree to which an innovation is perceived as being consistent with the existing values, past experiences, and needs of potential adopters.

Complexity: the degree to which an innovation is perceived as difficult to understand and use.

Diffusion: the process by which an innovation is communicated through certain channels over time among the members of a social system.

Early adopters: members of a group who have the greatest degree of opinion leadership and interpersonal networks; they are looked to by potential adopters for advice and information about the innovation.

Early majority: those who adopt new ideas just before the average member of a system; they provide interconnectedness in the system's interpersonal networks.

Innovation: an idea, practice, or object that is perceived as new by an individual or other unit of adoption (Rogers, 1995).

Innovation-decision process: the process through which an individual or group passes from first knowledge of an innovation to forming an attitude toward the innovation, to a decision to adopt or reject, to implementation and use of the new idea, and to confirmation of this decision(Rogers, 1995).

Innovators: members of a group who are venturesome, risk takers, and gatekeepers who protect the interest of the group's flow of new ideas from outside boundaries.

Instructional Design: the systematic development of instructional specifications using learning and instructional theory to ensure the quality of instruction.

Instructional Technology: the theory and practice of design, development, utilization, management and evaluation of processes and resources for learning.

Laggards: members of a group who are last in a social system to adopt an innovation; extremely cautious in adopting innovations.

Late majority: members of a group who adopt new ideas just after the average member of a system.

Observability: the degree to which the results of an innovation are visible to others.

Opinion Leaders: individuals who lead in influencing others' opinions about innovations.

Perception: one's reaction toward an object, idea, or situation; viewpoints or perspectives of individuals.

Relative advantage: the degree to which an innovation is perceived as better than the idea it supersedes.

Selective exposure: the tendency to attend to communication messages that are consistent with one's existing attitudes and beliefs. (Rogers, 1995)

Selective perception: a tendency to interpret communication messages in terms of one's existing attitudes and beliefs. (Rogers, 1995)

Social Change: the process by which alteration occurs in the structure and function of a social system.

Trialability: the degree to which an innovation may be experimented with on a limited basis. (Rogers, 1995)

References

Baker, W. & Gloster, A. (1994). Moving towards the virtual university: A vision of technology in higher education. *Cause, 17* (2), 4-11.

Beede, M. A. & Burnett, D. J. (1994). Student services for the 21st century: Creating the student-centered environment. In D. Oblinger & S. Rush (Eds.), The compatible campus: Planning, designing, and implementing information technology in the academy (pp. 68-86). Bolton, MA: Anker.

Berner, R. T. (1993). An electronic journal for the 21st century. Interpersonal Computing and Technology 1 (3), 1-6. Retrieved November 12,1998, from http://www.berner/ipctv1n3/Listserv@guvm

Bialo, E. R. & Sivin-Kachala, J. (1996). The effectiveness of technology in schools: A summary of recent research. Washington, DC: Software Publishers Association.

Biner, P. M. & Dean, R. S. (1997). Re-assessing the role of student attitudes in the evaluation of distance education effectiveness. In F. Saba (Ed.) Defining Concepts in Distance Education, p. 17-19.

Bjork, A. (1991). Is technology-based learning effective? *Contemporary Education, 63*, (1), 6-14.

Blackwell, R. D., Engle, J. F, & Miniard, P. W. (1995). *Diffusion of innovations in consumer behavior*. Dryden Press.

Boser, R., Palmer, J., & Daugherty, M. (1998). Students attitudes toward technology in selected technology education programs. *Journal of Technology Education, 10* (1), 1-13. Retrieved June 4, 2000, from www.scholar.lib.vt.edu/ejournals/JTE/

Boster, F., G. Meyer, A. Roberto, and C. Inge. (2002). "A Report on the effect of the unitedstreaming application on educational performance." United Learning. August.

Burke, J.(1994). Education is new challenge and choice: Instructional technology by way of superhighway? *Leadership Abstracts, 7* (10), 1-2.

Carr, P. (1999). The measurement of resourcefulness intentions in the adult autonomous learner (Doctoral dissertation, The George Washington University, 1999).

Cartwright, G. P. (1998). Technology implications for data system. *Change* CEO Forum on Education and Technology (1997). School technology and readiness report: From pillars to progress. Washington, DC: CEO Forum.

Clark, R. E. (1983). The next decade of instruction technology research. *Educational Considerations, 10* (2), 33-35.

Croft, R. S. (1994). What is a computer in the classroom? A Deweyan philosophy for technology in education. *Journal of Educational Technology Systems, 22* (4), 301-308.Cutshall, T. (1999). Tom's definition of instructional technology. Retrieved June 4, 2000, from http://www.arches.uga.edu/~cutshall/tomitdef.html

Cutshall, T. (1999). Tom's definition of instructional Add technology. Retrieved June 4, 200, from http://www.arches.uga.edu/-cutshall/tomitdef.html

Daniel, J. S. (1997). Why universities need technology strategies. *Change 31*(4), 11-17.

DeGroot, M. (2002). Multimedia projectors: A key component in the classroomof the future. *T.H.E.Journal.* Retrieved April 2, 2003, from http://thejournal.com/magazine/vault/A4056.cfm

DeLoughry, T. J. (1996). Reaching a critical mass: Survey shows record number of professors use technology in their teaching. *The Chronicle of Higher Education 42* (20), A17-A20.

Denk, J., Martin, J., & Sarangarm, S. (1993). Not yet comfortable in the classroom: A study of academic computing at three land-grant universities. *Journal of Educational Technology Systems, 22*(1), 39-55.

Denning, P. (1996). Business designs for the new university. *EduCom Review, 31* (4), 21-30.

DeSieno, R. (1995). The faculty and digital technology. *EduCom Review, 30* (4), 46-48.

Diamond P. & Simonson, M. R. (1988). Film-makers and persuasive films: A study to determine how persuasive films are produced. In M. R.

Simonson, (Ed.) Proceedings of the Annual Convention of the Association for Educational Communication and Technology (pp. 200-212).

Dick, W. & Carey, L. (1996). *The systematic design of instruction.* New York, NY: HarperCollins.

Draude, B. & Brace, S. (1999). Assessing the impact of technology on teaching and learning: Student perspectives. Paper presented at Mid-South Instructional Technology Conference, Murfreesboro, TN, March 28-30, 1999. Retrieved March 18, 2002, from http://www.mtsu.edu/~itconf/proceed99/brace.htm

Driscoll, M. & Dick, W. (1999). New research paradigms in instructional technology: An inquiry. *Educational Technology Research and Development 47* (2), 19-38.

Ehrmann, S. C. (1995). Asking the right questions: What does research tell us about technology and higher learning? *Change 27* (2), 20-27.

Ehrmann, S. C. (1997). The student as co-investigator. The Flashlight Program.Retrieved April 20, 2002, from http://www.tltgroup.org/programs/studco.html

El-Khawas, E. & Knopp, L. (1996). Campus Trends 1996 (Higher Education Panel Rep. No. 86). Washington, DC: American Council on Education.

Ely, D. P. (1997). The field of educational technology: Update 1997. A dozen frequently asked questions. Syracuse, NY: ERIC Clearinghouse on Information and Technology. (ERIC Document Reproduction Service No. ED 413 889)

Ester, D. P. (1995) CAI, lecture, and student learning style: The differential effects of instrumental method, *Journal of Research on Computing in Education, 27* (4), 129-139.Esther (1998). Instructional technology: Definition of the field. Retrieved April 20, 2002, from AltaVista http://www.techweb.com/encyclopedia/

Fife-Schaw, C., Breakwell, G. M., Lee, T. & Spencer, J. (1987). Attitudes towards new technology in relation to scientific orientation at school: A preliminary study of undergraduates, *British Journal of Educational Psychology, 57*, 114-121.

Fitzelle, G. T. & Trochim, W. M. (1996). Survey evaluation of web site instructional technology: Does it increase student learning? Retrieved

July 22, 1998, from http://www.trochim.human.cornell.edu/WebEval/webques/webques.htm

Geoghegan, W. H. (1994). Whatever happened to instructional technology? Paper presented at the 22nd Annual Conference of the International Business Schools Computing Association. Baltimore, MD, July 17-20, 1994. Retrieved June 2, 1999, from http://www.med.ibm.com/news/wmtep/whg/wpi.htm

Gentry, C. G. (1995). Educational technology: A question of meaning. In G. J. Anglin (Ed.), *Instructional technology. past, present, and future.* (pp. 1-7), Englewood, CO: Libraries Unlimited, Inc.

Geoghegan, W. H. (1998). Instructional technology and the mainstream: The risks of success. In D. Oblinger & S. Rush (Eds.), The compatible campus: planning, designing, and implementing information technology in the academy (pp. 131-150). Bolton, MA: Anker.

Gilbert, S. (1995). An online experience: Discussion group debates why faculty use or resist technology. *Change, 27* (2), 28-45.

Green, K. C. (1996). The coming ubiquity of information technology. *Change 28* (2), 24-31.

Green, K. C. (1997). Campus Computing 1996: The 1997 national survey of information technology in higher education. Encino, CA: Campus Computing. Retrieved April 19, 2001, from http://ericir@syr.edu/projects/campuscomputing/index

Green, K. C. (1998). *Campus Computing 1997.* Encino, CA: Campus Computing.

Green, K. C. (1999). *Campus Computing 1998.* The ninth national survey of desktop computing and information technology in American higher education. Encino, CA: Campus Computing.

Green, K. C., & Gilbert, S. W. (1995). Great Expectations: Content, communications, productivity, and the role of information technology in higher education. *Change, 27*(2), 8-18.

Gustafson, K. L. (1991). Survey of instructional development models (2nd ed.). Syracuse, NY: Educational technology Publications, (ERIC Document Reproduction Service No. ED 335-027)

Hahn, K. L. & Schoch, N. A. (1997). Applying diffusion theory to electronic publishing: A conceptual framework for examining issues and outcomes.

Retrieved August 13, 2002, from http://www.asis.org/annual-97/hahnk.htm

Hall, G. E., George, A. & Rutherford, W. (1979). *Measuring stages of concern about the innovation: A manual for the use of the social change questionnaire.* (Report No. 3032). Austin: The University of Texas at Austin, Research and Development Center of Teacher Education (ERIC Document Reproduction Service No. ED 147 342).

Holloway, R. E. (1996). Diffusion and adoption of educational technology: A critique of research design. In D. Jonassen (Ed.), *Handbook of research for educational communications and technology* (pp. 1107-1133). New York, NY: Simon & Schuster Macmillan.

Imel, S. (1998). *Technology and adult learning: Current perspectives* Columbus, OH: ERIC Clearinghouse on Adult Career and Vocational Education. (ERIC Document Reproduction Service No. ED 421 639)

Jacobsen, D. M. (1998). Adoption patterns and characteristics of faculty who integrate computer technology for teaching and learning in higher education. Unpublished doctoral dissertation, The University of Calgary, Canada.

Johnson, D. J. & Liu, L. (1998). Considering students' attitudes about information technology in planning education courses. Unpublished article, University of Nevada, Reno, NV.

Johnson, J. (1989). Effects of successful female role models on young women's attitudes toward traditionally male careers. In M.R. Simonson (Ed.) Proceedings of the Annual Convention of the Association for Educational Communication and Technology.

Kearsley, G. (1993, Spring/Summer). Educational technology: Does it work? *Ed-Tech Review*, 34-36.

Kerr, S. (1991). Lever and fulcrum: Educational technology in teachers' thought and practice. *Teachers College Record, 93*, 36-49.

Kershaw, A. (1996). People, planning, and process: The acceptance of technological innovation in post-secondary organizations. *Educational Technology, 36* (5), 44-48.

Knapper, C. K. (1980). *Evaluating instructional technology*, New York: Wiley.

Knezevich, S. J. & Eye, G. G. (1970). Instructional technology and the school administrator. Washington, DC: American Association of School Administrators.

Kosakowski, J. (1998). The benefits of information technology Syracuse, NY: ERIC Clearinghouse on Information and Technology. (ERIC Document Reproduction Service No. ED 420 302)

Kuh, G. D., and Vesper, N. (2001). Do computers enhance or detract from student learning? *Research in Higher Education*, 42 (1), pp. 87-101.

Lambert, R. (2000). What is instructional technology? Instructional Technology and Web management, Community College of Baltimore County (CCBC). Retrieved March 9, 2001, from http://www.ccbc.cc.md.us/ccbc/itwm/itwm.htm

Marien, M. (1996). New communications technology: A survey of impacts and issues. *Telecommunications Policy, 20* (5), 375-387.

Mayer, K. R. (2000). Student attitudes toward instructional technology in the large introductory U.S. Government course. American Political Science Association in Association with The Gale Group and LookSmart. Retrieved March 24, 2002, from http://www.findartcles.com

Mesher, D. (1999). Designing interactivities for internet learning. *Syllabus,12* (7), 16-20.

Moore, G. A. (1991). *Crossing the chasm.* New York: HarperBusiness.

Moore, M. (1997). Distance learning: A systems approach, Belmont: Wadsworth.

Myers, D. (1998). *Psychology,* New York, NY: Worth Publishers.

Pea, R. D. & Soloway, E. (1987). Mechanisms for facilitating a vital and dynamic education system: Fundamental roles for education science and technology. In B. Branyan-Broadbent & R. K. Woods (Eds.), *Educational Media and Technology Yearbook*, 1990, (pp. 31-90). Englewood, CO: Libraries Unlimited, Inc.

Perelman, L. (1993). *Schools out: Hyperlearning, thenew technology and the end of education.* New York, NY: Avon.

Ponton, M. (1999). The measurement of an adult's intention to exhibit personal initiative in autonomous learning (Doctoral dissertation, The George Washington University, 1999).

Rakes, G. C. (1996). Using the internet as a tool in a resource-based learning environment. *Educational Technology, 36* (3), 52-56.

Redish, E., Steinberg, R., & Saul, J. (1996). The distribution and change of student expectations in introductory physics. Department of Physics, University of Maryland, College Park. Paper presented at The International Conference on Undergraduate Physics Education, College Park, MD: The American Institute of Physics. Reeves, T. (1995). Questioning the questions of instructional technology research. Retrieved from http://www.hbg.psu.edu/bsed/intro/docs/dean/

Reed,R.(2003).Streaming technology improves student achievement *T.H.E. Journal.* Retrieved April 2, 2003, from http://www.thejournal.com/magazine/vault/A4320.cfm

Reigeluth, C. M. (1996). A new paradigm of ISD? Educational Technology, 36 (3), 13-20.

Reigeluth, C. M. (1997). Instructional theory, practitioner needs, and new directions: Some reflections. *Educational Technology, 37*(1),42-47.

Rogers, E. M. (1986). Communication technology: The new media in society. New York: The Free Press.

Rogers, E. M. (1995). *Diffusion of innovations* (4th ed.). New York: The Free Press.

Saba, F. (1998, January). Faculty and distance education. *Distance Education Report, 2* (1), 2-5.

Saettler, P. E. (1968). *A history of instructional technology*. New York, NY: McGraw-Hill.

Saettler, P. E. (1990). *The evolution of American educational technology.* Englewood, CO: Libraries Unlimited.

Santhanam, E. & Leach, C. (2000). University students' perceptions of information technology. In A. Herrmann & M. M. Kulski (Eds.), Flexible futures in tertiary teaching. Proceedings of the 9th Annual Teaching Learning Forum, February 2-4, 2000. Retrieved May 18, 2002, from http://cleo.murdoch.edu.au/confs/tlf/tlf2000/santhanam1.html

Schuyler, G. (1998). *A paradigm shift from instruction to learning* (ERIC Document Reproduction Service No. ED 414 961)

Seels, B. (1997). Taxonomic issues and the development of theory in instructional technology. *Educational Technology, 37* (1), 12-21.

Seels, B. B. & Richey, R. C. (1994). Instructional technology: The definition and domains of the field. Washington, DC: Association for Educational

Communications and Technology. (ERIC Document Reproduction Service No. ED 413 889)

Shields, M. (1995). Academe enters the age of anticippointment. *Technos, 4* (3). 30-32.

Simonson, M. (1995). Instructional technology and attitude change. In G. J. Anglin (Ed.), *Instructional technology: Past, present and future* (pp. 365-373). Englewood, CO: Libraries Unlimited.

Simonson, M. & Maushak, N. (1996). Instructional technology and attitude change. In D. H. Jonassen (Ed.), *Handbook of research for educational communications and technology* (pp. 984-1015). New York, NY: Simon & Schuster Macmillan.

Spotts, T. H., & Bowman, M. (1995). Faculty use of instructional technologies in higher education. *Educational Technology, 35* (2), 56-64.

Surry, D. W. (1997). Diffusion theory and instructional technology. Retrieved July 21, 1998, from http://www.hbg.psu.edu/bsed/intro/docs/diffusion/index

Surry, D. W. & Farquhar, J. D. (1996). Incorporating social factors into instructional design theory. Retrieved July 28, 1998, from http://www.hbg.psu.edu/bsed/intro/docs/social/August

Surry, D. W. & Gustafson, K.L. (1994). The role of perceptions in the adoption of computer-based learning. (ERIC Doc. Reprod. Service No. ED 374 788)

Twigg, C. (1994). The changing definition of learning. *EduCom Review*, 29(40), 23-25.

Twigg, C. A. (2000). The Pew learning and Technology Newsletter 2, (4), December 2000. Retrieved January 2, 2001, from http://PLTP-L@lists.rpi.edu

Vanderbilt Cognition and Technology Group. (1990). Anchored instruction and its relationship to situation learning. *Educational Researcher*, 2-10.Van Dusen, G. C. (1997). *The virtual campus: Technology and reform in higher education*, ASHE-ERIC Higher Education Report, 25 (5), Washington, DC: The George Washington University, School of Education and Human Development (ERIC Document Reproduction Service No. ED 412 815)

Wellburn, E. (1996). The status of technology in the education system: A literature review. Retrieved June 3, 1997, from http://www.cln.org/lists/nuggets/EdTechreport.html

Whitehead, B. M., Jensen, D., and Boschee, F. (2003). Planning for technology: A guide for school administrators, teacher coordinators, and curriculum leaders, Thousand Oaks, CA: Corwin Press.

Wilson, F. (1992). Language, technology, gender, and power. *Human Relations, 45*, 883-904.

Winner, L. (1995). The virtually educated. *Technos, 3* (4), 10-11.

Zimbardo, P. & Leippe, M. (1991). *The psychology of attitude change and social influence.* Philadelphia, PA: Temple University Press.

Appendix A

Students' Attitudes, Perceptions, and Expectations Toward Technology(SAPETT) Questionnaire

1. I feel that I the instructional technology course will provide me with an interactive learning experience with other classes.
2. I like group discussions in a new learning environment.
3. I believe that instructional technology will help to enhance my learning experiences in class.
4. I feel that the electronic classroom environment will provide a positive learning experience for me.
5. I feel instructional technology is a convenient means of acquiring new knowledge.
6. I think this classroom environment will help improve my grades.
7. I like this course because there are no set restrictions on my response time to class assignments.
8. I think this class will be a good learning experience for me.
9. I enjoy coming to this class because I like learning new things.
10. I feel that the new learning environment in the instructional technology class will not restrict my learning abilities.
11. I feel that I am learning more from this class.
12. For the courses I am enrolled in this spring 2001 semester, I send e-mail communication to fellow students.
13. For the courses I am enrolled in this spring 2001 semester, I send e-mail communication to instructors/professors.
14. I feel that writing e-mail messages will improve my writing skills.
15. I am able to work effectively in groups when completing course assignments.
16. I am able to initiate effective working relationships with fellow students when studying for exams.
17. I am able to have access to a computer off campus.
18. I am able to connect to the university's network from off campus.
19. I am required to learn in a seminar setting environment.

20. I think this introductory class of instructional technology should be a required course.
21. I feel that the new learning environment in the instructional technology class will not restrict my learning abilities.
22. I want an integrated learning environment with access to other knowledge outside the classroom environment.
23. Instructional technology is a tool for me to have group discussions with other students in other classes.
24. I see technology as an unlimited source for gathering information for my use as well as sharing information with others anywhere and at any time in the world.
25. I believe instructional technology is an expensive way to continue my education.
26. Taking classes in instructional technology will enable me to become a better student.
27. The introduction to instructional technology course is helping me to make use of my time.
28. For the courses I am enrolled in this spring 2001 semester, I have access to course materials through the Internet.
29. I am required to learn in a large lecture environment.
30. With the use of instructional technology, I expect the instructor to use other technology tools to broaden my knowledge.
31. With the use of instructional technology, I expect to be able to communicate with my instructors.
32. I am satisfied with the instructions I have received from the instructor/ professor regarding the process of the course.
33. I am satisfied with the instructor's method of teaching.
34. I am satisfied with the instructor's fairness in grading my work.
35. I am satisfied with the overall instructional environment.
36. I am satisfied with the expectations of the instructor.
37. I am satisfied with the expectations I have of the introduction to instructional technology class.
38. I expect this course will enable me to be prepared for future job opportunities.
39. The class is everything I expected and more.
40. I would recommend this course to other students.

Appendix B

Students' Attitudes, Perceptions, and Expectations Toward Technology (SAPETT) Survey

Name of your College/University: ______________________________

Today's date: ____________ Last four digits of your ID#: __________
(month) (day) (year) (social security #)

DIRECTIONS: This questionnaire is intended to elicit your opinion about how you are adapting to changes in your learning environment associated with an introductory course of instructional technology in higher education. Place an **X** on the line to show **how often the item applies to you**. Your mark can be on any spot on the scale from 0 to 10.

A score of **0** means that the statement is **never** descriptive of me. A score of **5** means that the statement is **sometimes** descriptive of me. A score of **10** means that the statement is **always** descriptive of me. For example, if you wanted to respond to sample question with a score of 2.75, then you would mark the answer sheet as follows:

I will like my learning experiences in the instructional technology class.

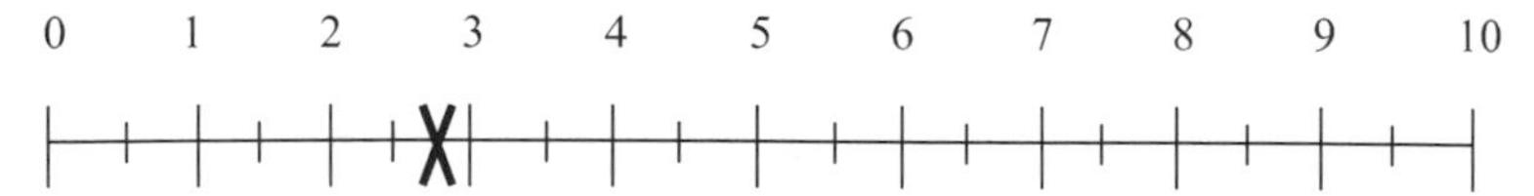

1. I feel that the instructional technology course will provide me with an interactive learning experience with other classes.

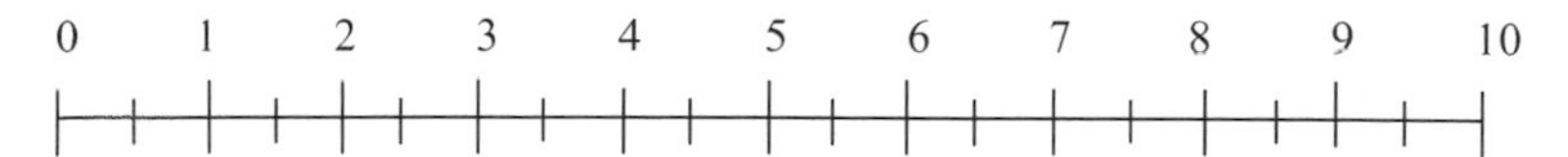

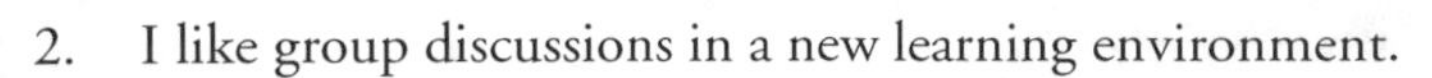

2. I like group discussions in a new learning environment.

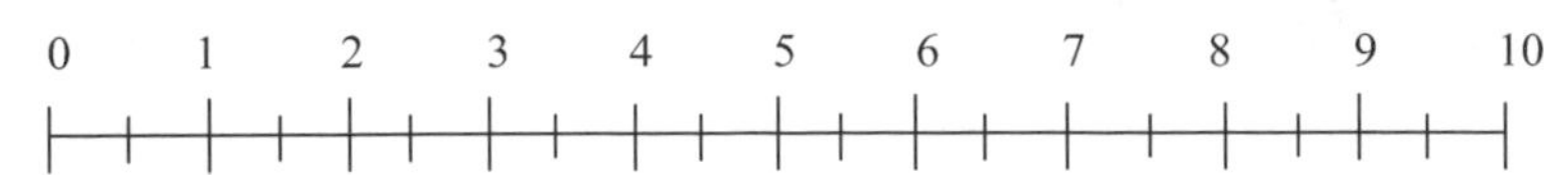

3. I believe that instructional technology will help to enhance my learning experiences in class.

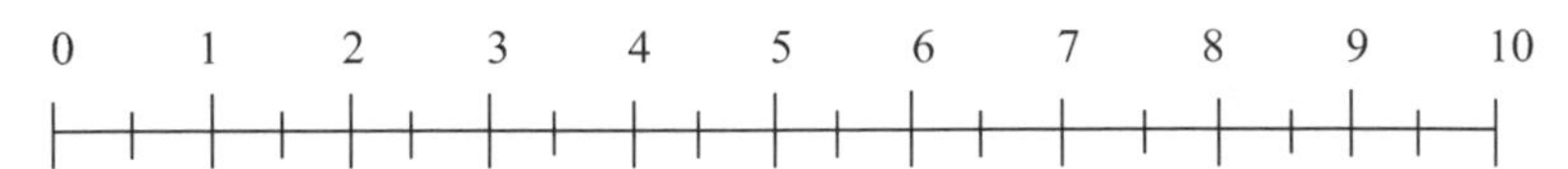

4. I feel that the electronic classroom environment will provide a positive learning experience for me.

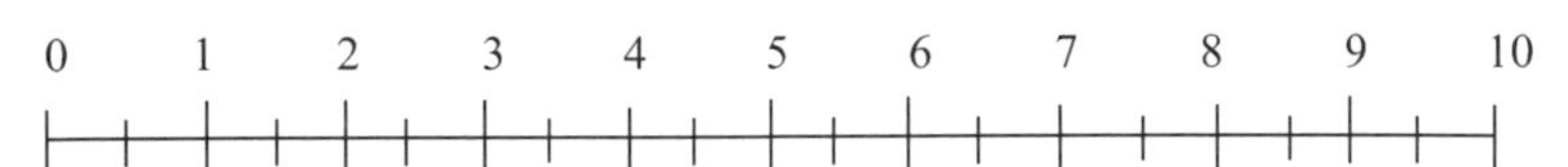

5. I feel instructional technology is a convenient means of acquiring new knowledge.

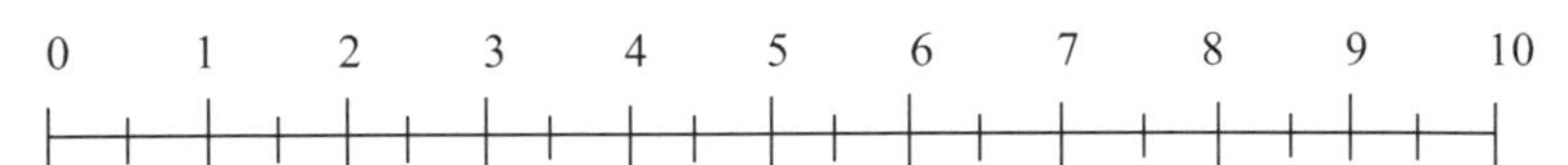

6. I think this classroom environment will help improve my grades.

0 1 2 3 4 5 6 7 8 9 10

7. I like this course because there are no set restrictions on my response time to class assignments.

0 1 2 3 4 5 6 7 8 9 10

8. I think this class will be a good learning experience for me.

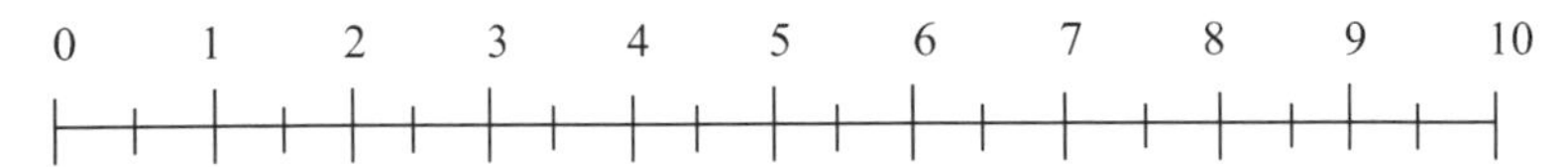

9. I enjoy coming to this class because I like learning new things.

0 1 2 3 4 5 6 7 8 9 10

10. I feel that the new learning environment in the instructional technology class will not restrict my learning abilities.

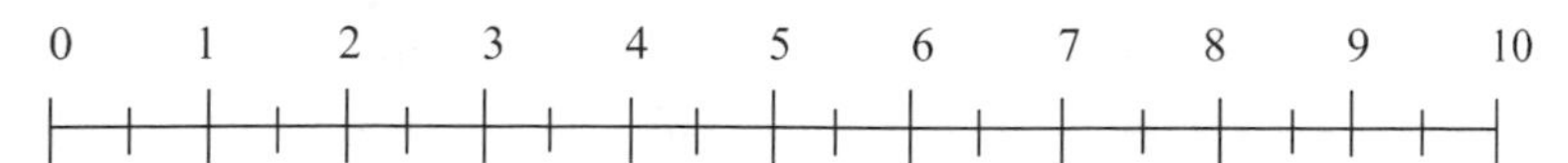

11. I feel that I am learning more from this class.

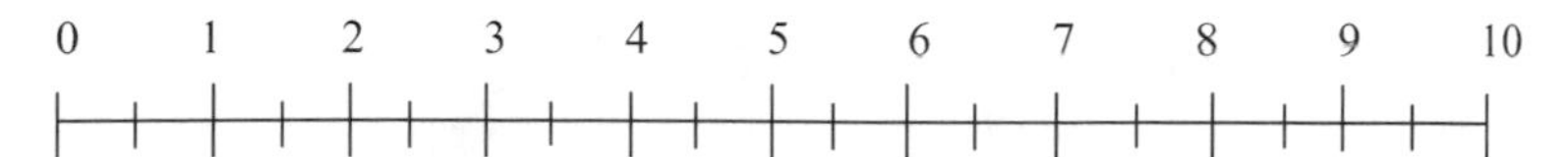

12. For the courses I am enrolled in this Spring 2000 semester, I send e-mail communication to fellow students.

0 1 2 3 4 5 6 7 8 9 10

13. For the courses I am enrolled in this Spring 2000 semester, I send e-mail communication to instructors/professors.

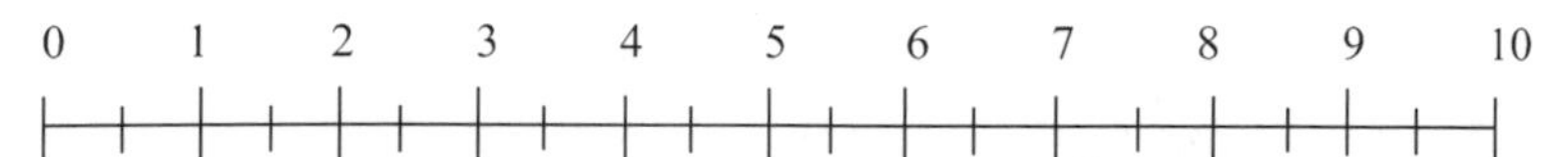

14. I feel that writing e-mail messages will improve my writing skills.

0 1 2 3 4 5 6 7 8 9 10

15. I am able to work effectively in groups when completing course assignments.

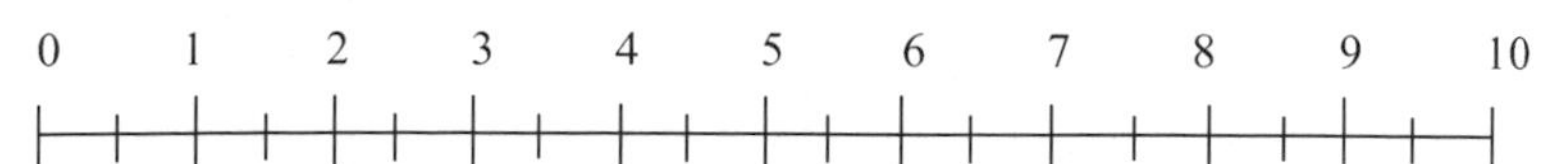

16. I am able to initiate effective working relationships with fellow students when studying for exams.

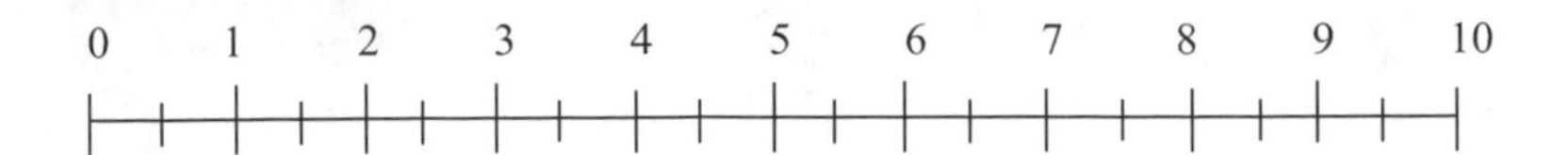

17. I am able to have access to a computer off campus.

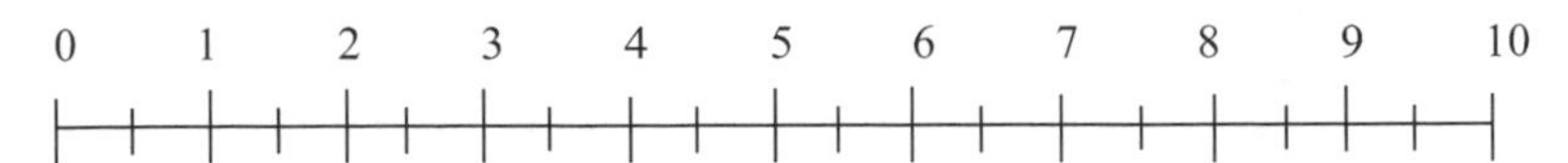

18. I am able to connect to the university's network from off campus.

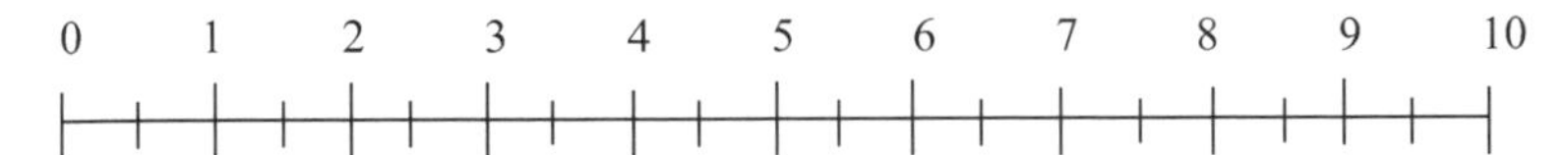

19. I am required to learn in a seminar setting environment.

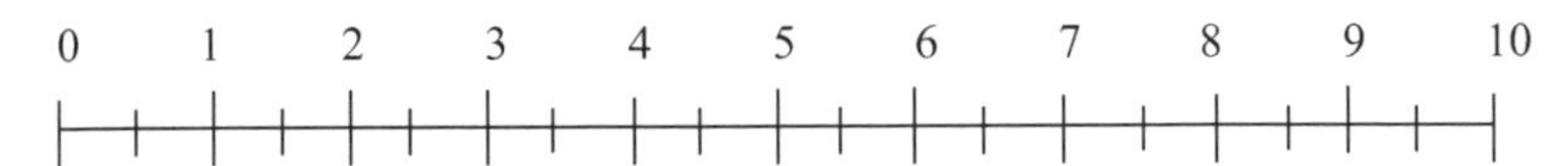

20. I think this introductory class of instructional technology should be a required course.

0 1 2 3 4 5 6 7 8 9 10

21. I feel that the new learning environment in the instructional technology class will not restrict my learning abilities.

0 1 2 3 4 5 6 7 8 9 10

22. I want an integrated learning environment with access to other knowledge outside the classroom environment.

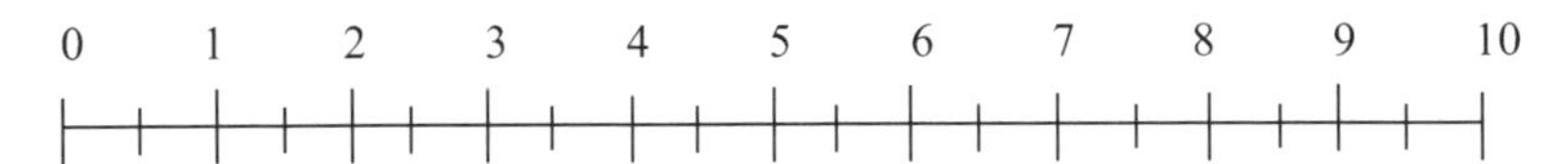

23. Instructional technology is a tool for me to have group discussions with other students in other classes.

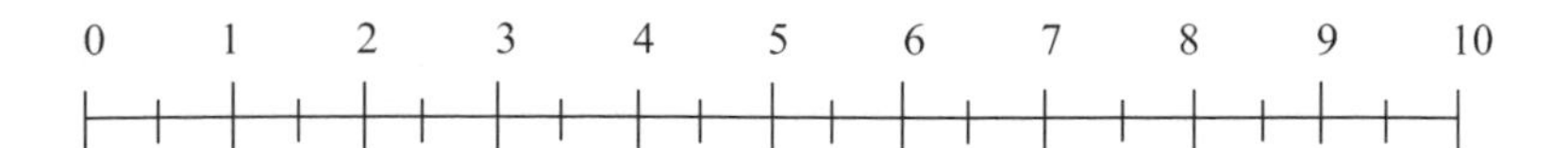

24. I see technology as an unlimited source for gathering information for my use as well as sharing information with others anywhere and at any time in the world.

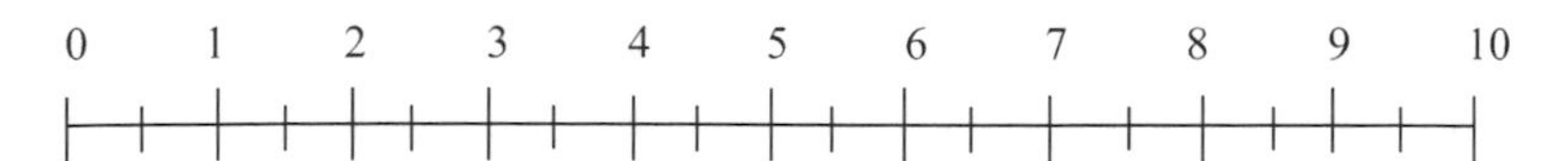

25. I believe instructional technology is an expensive way to continue my education.

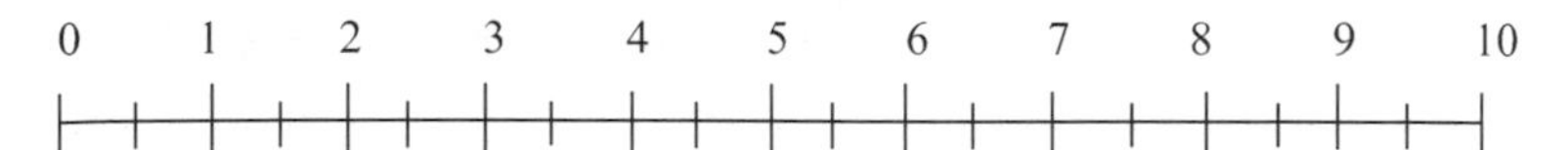

26. Taking classes in instructional technology will enable me to become a better student.

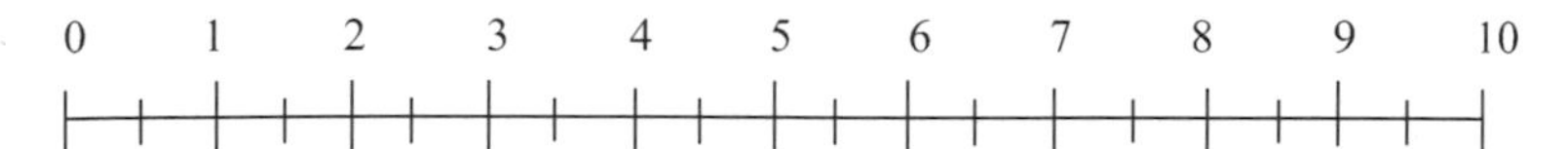

27. The introduction to instructional technology course is helping me to make use of my time.

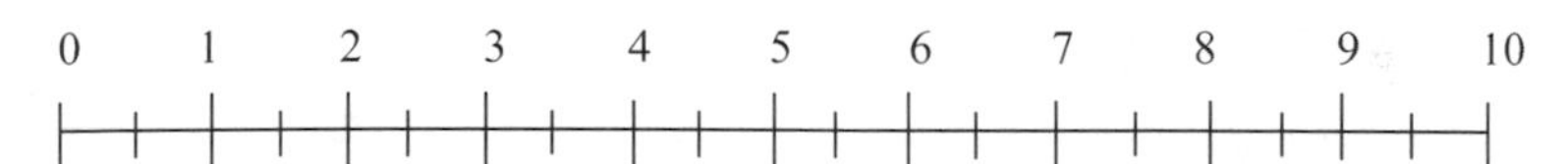

28. For the courses I am enrolled in this Spring 2000 semester, I have access to course materials through the Internet.

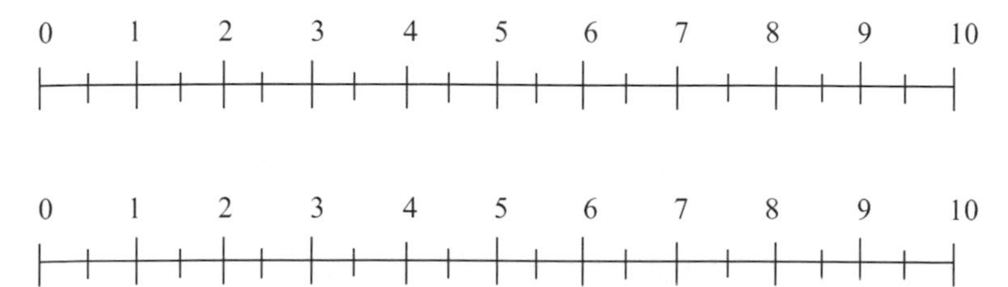

29. I am required to learn in a large lecture environment.

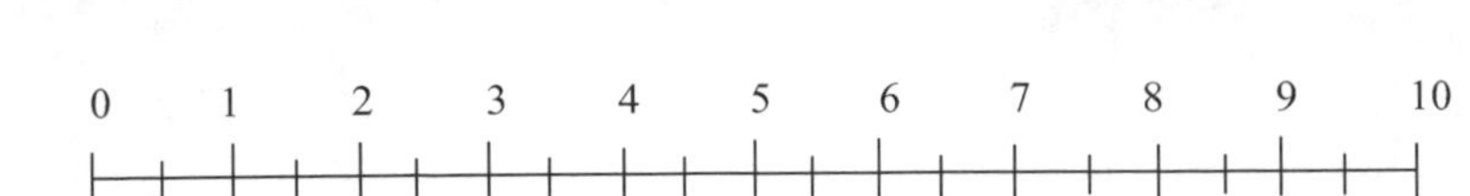

30. With the use of instructional technology, I expcct the instructor to use other technology tools to broaden my knowledge.

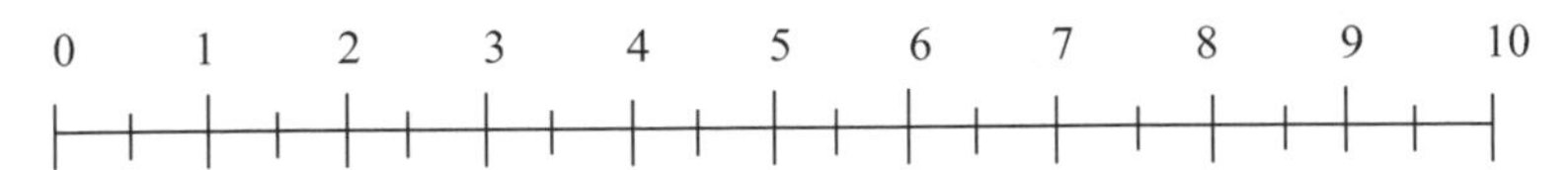

31. With the use of instructional technology, I expect to be able to communicate with my instructors.

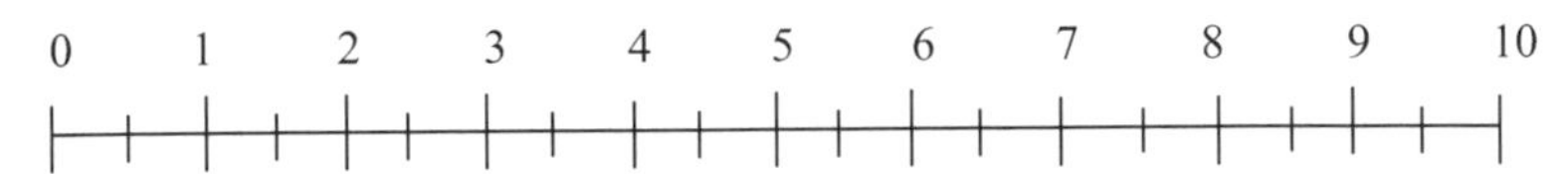

32. I am satisfied with the instructions I have received from the instructor/ Professor regarding the process of the course.

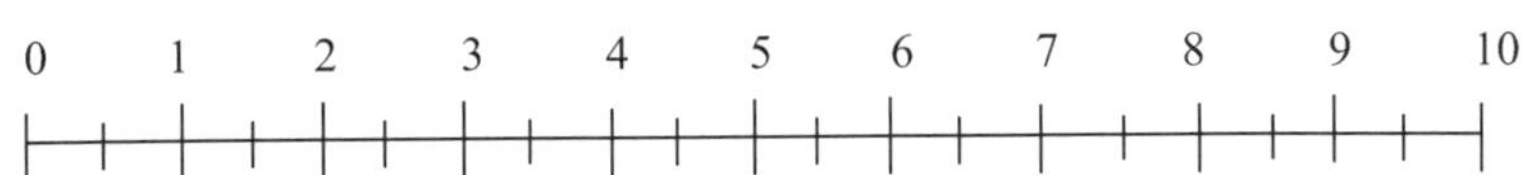

33. I am satisfied with the instructor's method of teaching.

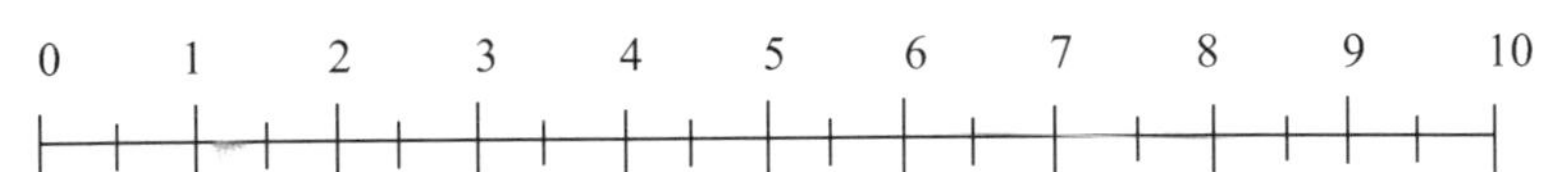

34. I am satisfied with the instructor's fairness in grading my work.

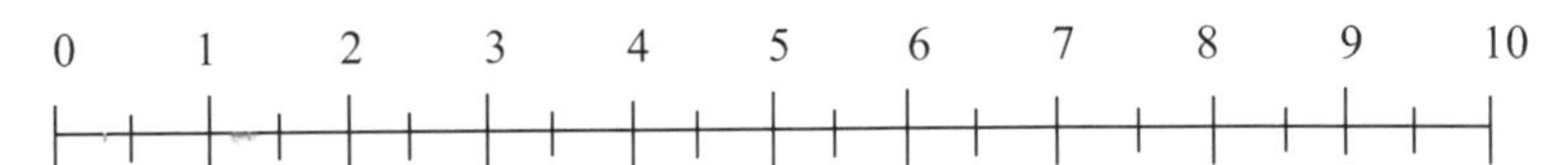

35. I am satisfied with the overall instructional environment.

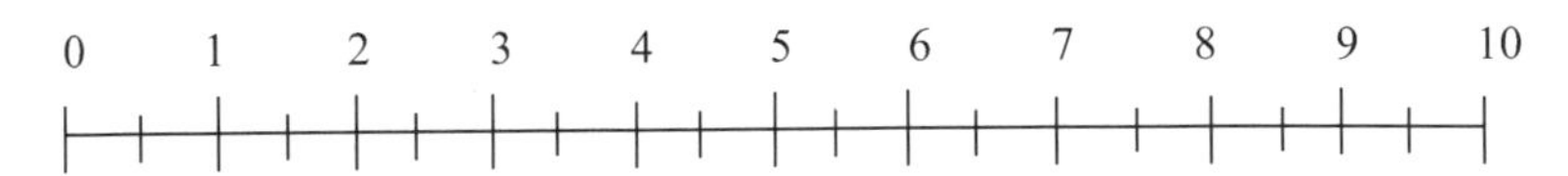

36. I am satisfied with the expectations of the instructor.

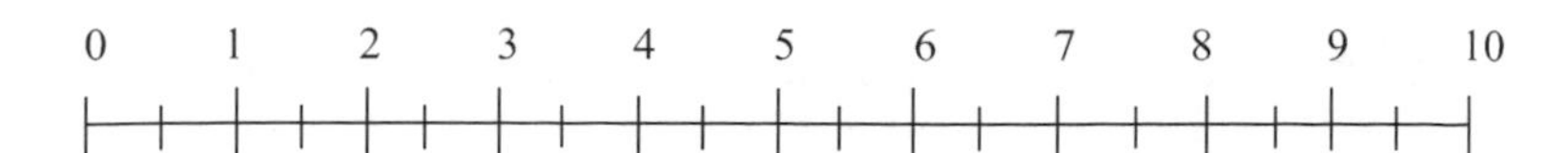

37. I am satisfied with the expectations I have of the introduction to instructional technology class.

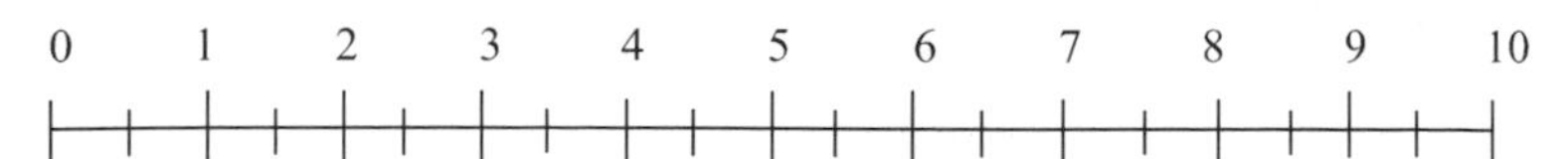

38. I expect that this course will enable to be prepared for future job opportunities.

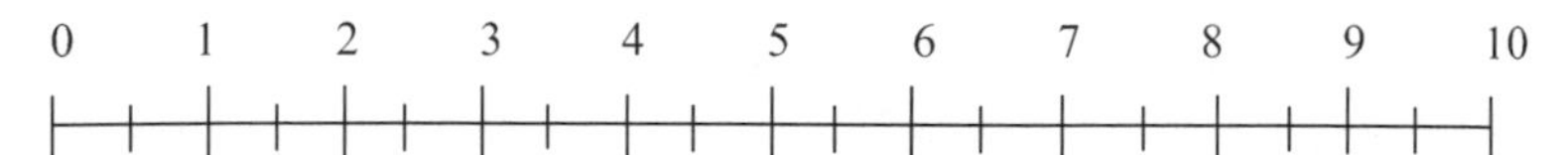

39. The class is everything I expected and more.

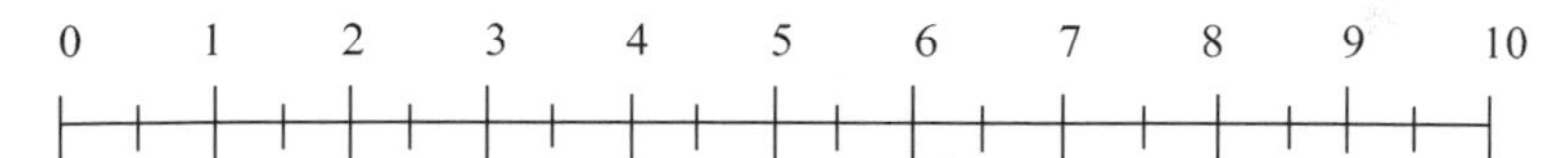

40. I would recommend this course to other students.

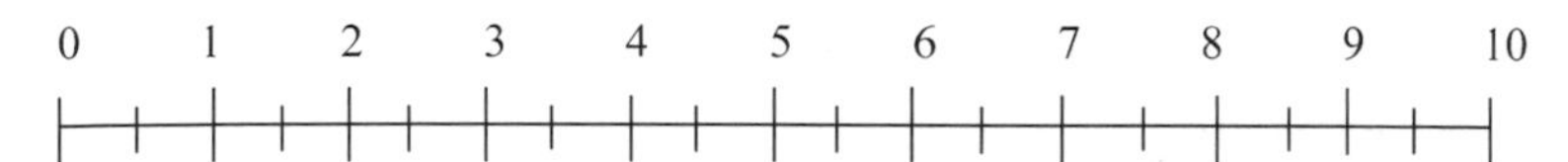

DEMOGRAPHICS:

Gender: Male ____ Female ____

Classification: Freshman ___ Sophomore ___ Junior ___ Senior ___

Annual family income:
(less than $25,000) ___ ($25,000-$69,999) ___ ($70,000 or more) ___

Age: ____ Exposure to Instructional Technology: ______________
(college credit)

Describe yourself in terms of ethnicity in one or two words:

____

____

Index

A

B

C

D

S

T

U

V

W

Z